THE COLLINS ENCYCLOPEDIA OF
ANIMAL
EVOLUTION

THE COLLINS ENCYCLOPEDIA OF
ANIMAL
EVOLUTION

Edited by Professor R. J. Berry and
Professor A. Hallam

COLLINS
8 Grafton Street, London W1

Project Editor: Graham Bateman
Editors: Robert Peberdy, Philip Gardner
Art Editor: Chris Munday
Art Assistant: Wayne Ford
Picture Research: Alison Renney
Production: Clive Sparling
Design: Adrian Hodgkins
Index: Barbara James

 AN EQUINOX BOOK

Published by:
William Collins Sons & Co Ltd,
London, Glasgow, Sydney, Auckland,
Toronto, Johannesburg

First published in 1986

Planned and produced by:
Equinox (Oxford) Ltd
Littlegate House
St Ebbe's Street
Oxford OX1 1SQ

ISBN 0-00-219818-5

Origination by Scantrans, Singapore

Filmset by BAS Printers Ltd,
Over Wallop, Stockbridge, Hampshire, England

Printed in Spain by Heraclio Fournier SA, Vitoria

Advisory Editors

Professor James W. Valentine
University of California,
Santa Barbara,
California, USA

Professor David B. Wake
University of California,
Berkeley,
California, USA

Artwork Panels

Michael R. Long

Graham Allen

Trevor Boyer

Richard Hook

Richard Lewington

Richard Orr

Denys Ovenden

Contributors

WA   Wallace Arthur
Sunderland Polytechnic
England

RJB   R. J. Berry
University College London
England

AJCa   A. J. Cain
University of Liverpool
England

PC   P. Calow
University of Sheffield
England

HLC   Hampton L. Carson
University of Hawaii
Honolulu
Hawaii, USA

AJC   Alan J. Charig
British Museum (Natural History)
London
England

ENKC   Euan N. K. Clarkson
Grant Institute of Geology
University of Edinburgh
Scotland

SCM   Simon Conway Morris
University of Cambridge
England

LMC   Lawrence M. Cook
University of Manchester
England

RHC   Robert H. Cowie
University College London
England

RAF   Richard A. Fortey
British Museum (Natural History)
London
England

AH   Anthony Hallam
University of Birmingham
England

GAH   G. Ainsworth Harrison
University of Oxford
England

JRM   James R. Moore
The Open University
Milton Keynes
England

PDM   Peter D. Moore
King's College London
England

DTP   David T. Parkin
University of Nottingham
England

LPa   Linda Partridge
University of Edinburgh
Scotland

PWSk   Peter W. Skelton
The Open Univeristy
Milton Keynes
England

PJBS   Peter J. B. Slater
University of St Andrews
Scotland

SMS   Steven M. Stanley
Johns Hopkins University
Baltimore, Maryland
USA

AJS   Tony J. Stuart
University of Cambridge
England

Left: Hot bed of evolution. Life first evolved in the seas and these remain very complex ecosystems with a wide variety of life forms present, as this scene from the Red Sea illustrates (Oxford Scientific Films). Half-title: Charles Darwin, photographed just before his death (Mansell collection). Title-page: Endemic dinner—a Galapagos hawk (Buteo galapagoensis) eats a Marine iguana (Amblyrhynchus cristatus). Both these species are unique (endemic) to the Galapagos Islands, the visit to which by Charles Darwin was vital to his formulation of the theory of evolution by natural selection (Andrew Laurie).

PREFACE

T. H. Huxley, friend, supporter and protagonist of Charles Darwin, is reported to have exploded when Darwin's theory of evolution by means of natural selection was explained to him. "How extremely stupid not to have thought of that," he is supposed to have said. And many others have responded similarly when faced with Darwin's simple argument: the fact that there is inherited variation between individuals struggling for existence means that the possessors of some variants will be more likely to survive and reproduce than others. The result is that as time passes, an increasing proportion of individuals will be found with the particular advantageous variant. Reduced to its most basic level, that is all there is to Darwinian evolution. Darwin provided an easily understood and testable mechanism whereby the adaptation of animals and plants to their environment could take place. It is no wonder that evolution was so rapidly accepted after the publication of *On the Origin of Species* . . . in 1859. Evolution is the theoretical core to biology, just as the Periodic Table is to chemistry.

Why then, does evolution attract controversy and dissent? Why do scientists and "creationists" argue so vehemently about it? There are at least three different reasons:

(1) **Evolution is a fact.** The fossil record clearly shows a progression from simpler to more complex forms over millions of years. There is also a multiplicity of evidence for descent from a common ancestor from studies in comparative anatomy, embryology, molecular biology and much more. There are gaps in our detailed knowledge of this history, but there is no serious debate about its occurrence. Where there is legitimate debate it is over the mechanisms by which evolutionary change occurs. There are ways by which genetic change can take place by mechanisms other than natural selection and scientists argue about the relative importance of these, as they do about the rates of evolutionary change. Such arguments are the legitimate business of science, leading to hypothesis, experiment, observation and conclusion. They should not be confused with doubts about whether or not evolution has occurred.

(2) **The role of mutations.** It is frequently claimed that the evolution of so-called "perfect" organs like the human eye or ear, or of the intricate interplay of species in the natural world, could not possibly arise from random, usually deleterious, alterations (mutations) in genes. This misconception arises from the assumption that evolution is driven by mutation; put another way, that evolutionary change depends solely on rare advantageous mutations. As we shall see in the following pages, virtually all populations of animals and plants have a wealth of inherited variations, and this pool of variation is available for selection. Mutation is the ultimate raw material of evolution, but evolutionary changes rarely depend on the fortuitous occurrence of new mutations at any place or time.

Over-stressing the importance of mutations is one example of the problem of looking at evolution from a particular point of view. The *Origin of Species* succeeded because Darwin himself was extremely adept at drawing together evidence from a wide variety of disciplines. Controversies about evolution almost inevitably break out when specialists look at it from their own limited perspective

(3) **Human beings are animals,** and are fairly closely related to the higher apes (chimpanzees, gibbons, gorillas, orang utans). Many people have a revulsion about this. Benjamin Disraeli, British Prime Minister in Queen Victoria's time asked, "Is man an ape or an angel? I am on the side of the angels." In fact, there is no reason why men and women should not be regarded as *both* animals *and also* human individuals existing and operating on a higher plane. Alfred Russel Wallace, co-discoverer with Darwin of natural selection, held this belief, and many distinguished scientists have had a similar faith. In more general terms, there is no inevitable conflict between evolution and religious faith.

Evolution is change, and is therefore a challenge and perhaps a threat to an organism. Perhaps the recurring popular arguments about evolution can be regarded as stemming from a subconscious fear that we will be similarly challenged or threatened ourselves. Whether or not this is so, it is fair to note that there are legitimate scientific debates about evolution, and the more important ones are described in this book. But these are debates about the *mechanism(s)* of evolutionary change, not about the occurrence of evolution itself.

This book is divided into six sections, each section comprising one or more entries. In the first section we look at The Prehistoric World and see what animals have existed in the constantly changing environment of this planet. Also the discipline of paleontology (the study of fossils) is introduced.

The Background to Evolution not only reviews the history of Charles Darwin and the "evolution" of the theory of natural selection, but also investigates earlier theories as far back as the times of Aristotle and Cicero. This is a fascinating story, continually interwoven with theological and political views and principles, and rivalry between natural historians. Having accepted the validity of the theory of evolution, the third section on The Course of Evolution assesses the evidence that evolution has occurred, with particular emphasis on the fossil record.

The Consequences of Evolution looks in detail at the results of the fact that all living things have a common ancestry— dating as far back as the origin of life itself. Evolution is about changes and relationships; the living world is not static, new species arise while others vanish. But how does this come about? The Mechanisms of Evolution attempts to answer this question.

The final section deals with two aspects of the human story. Firstly it investigates the evolution of the human species on this planet. And secondly it summarizes the arguments about evolution—both scientific and also ones about where God fits into an evolving world.

Each entry is introduced by a summary of its contents, which is followed by the main text. Subjects which require more detailed coverage or which provide a wider perspective are covered in boxed features or double-page features.

The entries have been written by a team of 21 authors, all renowned scientists of considerable international reputation. Clearly a volume encompassing life in the past cannot be totally illustrated with photographs. We have been fortunate to assemble a team of artists who have brought to the book their interpretations of the views of the authors on what life was like in the past. To these have been added a liberal lacing of diagrams, photographs and portraits, captions to which frequently give more detailed background information. The whole project has been arranged and coordinated by the editorial team of Equinox (Oxford) Limited.

Let Darwin have the last word. He wrote the *Origin of Species* in a great hurry. At the end of the book he took breath and surveyed the whole world of nature which would ever after be viewed in a different way. The *Origin* ends:

"It is interesting to contemplate an entangled bank, clothed with many plants of many kinds, with birds singing on the bushes, with various insects flitting about, and with worms crawling through the damp earth, and to reflect that these elaborately constructed forms, so different from each other, and dependent on each other in so complex a manner, have all been produced by laws acting around us. These laws, taken in the largest sense, being Growth with Reproduction; Inheritance, which is almost implied by reproduction; Variability, from the indirect and direct action of the external conditions of life, and from use and disuse; a Ratio of Increase so high as to lead to a Struggle for Life, and as a consequence to Natural Selection, entailing Divergence of Character and the Extinction of less-improved forms. Thus, from the war of nature, from famine and death, the most exalted object which we are capable of conceiving, namely, the production of the higher animals, directly follows. There is a grandeur in this view of life, with its several powers, having been originally breathed by the Creator into a few forms or into one; and that, whilst this planet has gone cycling on according to the fixed law of gravity, from so simple a beginning endless forms most beautiful and most wonderful have been, and are being, evolved."

Sam Berry UNIVERSITY COLLEGE LONDON

Tony Hallam THE UNIVERSITY OF BIRMINGHAM

Last of the land giants—an African elephant intimidates a lion. The two elephant species are the last of a line of larger herbivores (order Proboscidea) for which the evolution of large body size and the ability to eat large quantities of less nutritious browse, removed them from competition with wild asses, zebras and rhinos (Premaphotos Wildlife).

CONTENTS

Fossil ammonites, Arnicoceras semicostatum, *in a sample of rock from the lower Jurassic period, found at Robin Hood's Bay, Yorkshire, England. Ammonites lived from 400 to 65 million years ago and became extinct in the Cretaceous period (Sinclair Stammers).*

The Prehistoric World

EVOLUTION is the name given to the changes that have taken place during the history of life on this planet, and to the mechanisms that explain how the different kinds of animals and plants came into existence, and how they are related to each other. The later parts of this book explain in detail how evolution is believed to have occurred; what evidence has led biologists to that conclusion; and the mechanisms by which new species have been formed.

It is impossible to appreciate the weight of evidence for evolution, and the power of evolutionary theory to explain a multitude of facts about the living world, without first acquiring an understanding of the history of the earth from its formation some 4,600 million years ago. This geological history is deduced partly from the rock strata themselves, some of which can be dated by radiometric means, and partly from the fossils the rocks contain.

The fossil story is a fascinating one. From the tenuous earliest evidences of life in the Precambrian era, through the successive geological periods, the variety of life-forms preserved as fossils is both astonishing and instructive. New species appear and disappear, sometimes (but not always) suddenly—or so it seems, though "sudden" on a geological time scale can mean over millions of years. From the fossil record, we can tell not only what the animals and plants of the past looked like, but even to some extent how they lived, for footprints, burrows and other clues to behavior have also sometimes been preserved in fossil form.

In the pages that follow, the development of animal life is traced from the earliest times to the present epoch. The dramatic changes from one age of the earth to the next can sometimes be explained by the tremendous climatic changes that took place, and by the geological upheavals resulting from the splitting up and movement of the great supercontinents and continents. But much is still uncertain and mysterious; much yet remains to be understood.

◄ **Representative animals of the late Mesozoic era** (Cretaceous period, 144–65 million years ago; see p28). (1) A *Pteranodon*, a giant flying reptile. It probably soared on hot air currents. Maximum wingspan about 5m (16ft). (2) *Tylosaurus*, a giant sea reptile of the family of mosasaurs (Mosasauridae). Its diet probably consisted mainly of crustaceans, eg ammonites. Maximum length about 10m (33ft). (3) *Elasmosaurus platyurus*, a species of long-necked plesiosaur from the late Cretaceous seas of North America. Long-necked plesiosaurs were expert at fishing. Maximum length about 13m (44ft). (4) *Chelonia archelonischyros*, a species of giant marine turtle. It had very strong, paddle-like limbs which consisted of fingers joined by webs of skin. Maximum length about 4m (13ft). (5) *Hesperornis regalis*, a species of flightless bird of the late Cretaceous. It had vestigial wings and its legs and feet were adapted for swimming. Maximum length about 1·5m (5ft). (6) *Ichthyornis*, a genus of tern-like bird: the earliest-known bird with a keeled breastbone for the support of flight muscles. Maximum length about 20cm (8in).

PALEONTOLOGY – The Study of Fossils

What is paleontology?. . . Fossils and trace fossils. . . The history of paleontology. . . The role of fossils in the geological time-scale. . . Geological eras and the fossils they left behind. . . Invertebrate paleontology. . . Vertebrate paleontology and evolutionary biology. . . The history of land plants and micropaleontology. . . Trace fossils

PALEONTOLOGY is the study of prehistoric life; of the animals and plants that lived in the prehistoric past. Most paleontological information is provided by fossils (derived from the Latin word *fossilis*, meaning "dug up" or "unearthed") and by the rocks that enclose them. The majority of fossils are the dead remains of actual organisms, and these are correctly termed "body fossils." But paleontology also deals with fossils resulting from the activity of animals (trace fossils, see p10), which similarly give evidence of past life, though less directly. Other than the study of trace fossils (ichnology), paleontology is normally conveniently divided into four categories: the studies of fossil invertebrates (invertebrate paleontology), of fossil vertebrates (vertebrate paleontology), of fossil plants (paleobotany), and of fossil microorganisms (micropaleontology).

Fossils are found mainly in a general group of rocks known as sedimentary rocks. These consist of weathered particles of other rocks that are transported, usually by water, and deposited elsewhere in distinct layers (strata). There are a number of mechanisms whereby fossil remains are preserved; they include petrification, casts, compressions and mummification.

The History of Paleontology

Fossils have been known from the earliest days of human history, and in some primitive societies have been used as cult objects. In some ancient societies the nature of fossils as organic remains was fully appreciated. The writings of Xenophanes, Pythagoras and Herodotus, who lived in Greece in the 6th and 5th centuries BC, record that fishes and shells found in rocks far inland were the remains of animals and that the sea had once extended that far in former times. Some Chinese texts of the 10th century AD draw attention to fossil fishes found high on a mountain; the writer understood their nature as organic remains. In the western world, however, fossils were generally regarded, even as late as the 17th century, merely as "little sporting miracles of nature," or as resulting from "some extraordinary plastic virtue latent in the earth." Or, of course, as some of today's "creationists" still see them, as "relics of the Deluge." Leonardo da Vinci (1452–1519), however, with his capacity for acute observation, spoke of the "ordered layers" in which the shells were found, which to him did not speak of an origin in the Deluge. Leonardo had some influence upon his contemporaries, such as the physician Girolamo Fracastoro (1478–1553), but this did not long persist after his death. During the 17th and 18th centuries some excellent descriptions of fossils began to accumulate, especially after the work of the Dane Nicolaus Steno (1638–86) and later of the English microscopist Robert Hooke (1635–1703), who not only knew that fossils were the remains of animals and plants, but thought that they might give some indication as to the former distribution of land masses and of climates.

In the late 18th and early 19th centuries, owing to the work of Linnaeus, Lamarck and Cuvier, the study of paleontology became finally established as a descriptive science of intrinsic biological value. Moreover, as a result of the early work of William Smith (1769–1839) fossils proved to be of inestimable practical value in the correlation of rocks, and in providing a relative time scale for geology. Once the distribution of fossils in time (stratigraphical sequence) had been worked out in general terms, as it was by the 1830s, it became possible to assign any fossil-bearing rock to its correct position in the time scale, and this can now be done with refined precision.

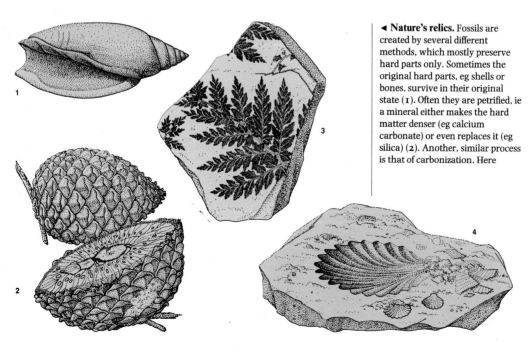

◄ **Nature's relics.** Fossils are created by several different methods, which mostly preserve hard parts only. Sometimes the original hard parts, eg shells or bones, survive in their original state (1). Often they are petrified, ie a mineral either makes the hard matter denser (eg calcium carbonate) or even replaces it (eg silica) (2). Another, similar process is that of carbonization. Here oxygen, hydrogen and nitrogen disappear, leaving behind concentrated carbon, usually as a flat image. This is called a compression (3). In some situations the animal's shell is worn away by water or another liquid and its shape is preserved as a mold. In this way either the external (4) or the internal form can be preserved.

► **Public awareness of fossils** ABOVE grew in the 19th century with the propagation of encyclopedic works.

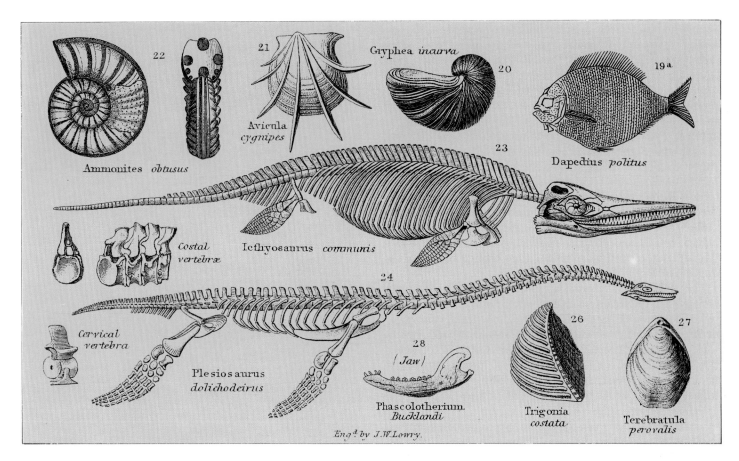

22

21

Gryphea *incurva*

20

19a

Avicula cygnipes

Ammonites *obtusus*

23

Dapedius *politus*

Costal vertebræ

Icthyosaurus *communis*

24

Cervical vertebra

26

27

28
(Jaw)

Plesiosaurus *dolichodeirus*

Phascolotherium *Bucklandi*

Trigonia *costata*

Terebratula *perovalis*

Eng.ᵈ by J.W.Lowry.

Fossils and the Geological Time Scale

This primary use of fossils in stratigraphy, however, is only possible if the species upon which such stratigraphy is based can be accurately identified and classified. Thus at the heart of all paleontological research stands taxonomy; that is, the description, naming and classification of animals and plants into ordered groups—species, genera, families and so on—which should reflect their evolutionary relationships. Species are described in technical detail and fully illustrated in paleontological papers and monographs. This is an essential requisite for all other uses of paleontology, so that the fossils may be identified with accuracy, and a great number of paleontologists are mainly concerned with the production of monographs.

Invertebrate fossils are common in rocks which were deposited during the last 600 million years, and the geological periods and systems from the Cambrian upward, together with the Paleozoic, Mesozoic and Cenozoic eras, were first defined in terms of the fossil faunas that they contain. Each of the systems has its characteristic fossil content, and within each system particular fossils of stratigraphic value define short time zones, which may represent periods of less than one million years.

Thus the whole geological column is divided up into short units based upon the time spans of particular fossil species. In the Paleozoic the most useful fossils for this purpose are trilobites, graptolites, conodonts and to a lesser extent brachiopods. In the Upper Paleozoic foraminiferids and early ammonoids are of most value, and although species of brachiopods and corals have longer time ranges, they have proved use-

ful at certain levels. In the Mesozoic, ammonoids are the best of all zone fossils, and in the Cenozoic microfossils and especially foraminiferids are of the greatest use. The systems are terminated by a significant change in the overall aspect of the marine faunas, in which many fossil groups become extinct at around the same time and are replaced by new ones. These extinction periods, of which by far the greatest was at the end of the Permian, are probably chiefly due to physical causes.

The relative time scale given by fossils is now enhanced by radiometric dates which can be determined from volcanic rocks interleaved within the fossil-bearing sedimentary sequences. Radiometric dating is a technique whereby the age of a rock or fossil can be determined by the comparison of the amount of a particular radioactive substance present there with the amount of one or more of its decay products. Knowing how quickly a particular element decays (its half-life) allows a precise calculation of the age of the rock. Hence the geological column, long subdivided into systems, is now properly dated in terms of millions of years. Thus for example the Cambrian period (the time during which the rocks of the Cambrian systems were laid down) lasted from about 590 to 505 million years ago.

Branches of Paleontology

Invertebrate paleontology is concerned with fossil animals without backbones. Invertebrates comprise about 95 percent of all animals. Most fossil invertebrates are found in marine sediments (those laid down by the sea); they are surprisingly common in various kinds of sedimentary rock, such as clays, limestone, shales, silts and sometimes sandstones. In such rocks it is normally only the hard shells that are preserved, and the soft parts, together with the remains of soft-bodied animals, normally decay without trace. Preservation is favored by rapid burial, and while the best-preserved marine fossils are commonly found in limestones or calcareous shales, preservation can be—for better, or more often for worse—affected by changes in the sediment after burial as it is compacted into rock (diagenesis). Sometimes, as is common in siltstones, the fossil shell is dissolved leaving molds of the internal and external surface in the enclosing rock. A band of such dissolved shells will form a distinct plane of weakness in the rock. When the rock is cracked open along this plane it reveals a fine array of fossil shells. The detailed structure of these can be studied by pouring a latex or silicone-rubber solution onto the surface; when it has set it is stripped off so that it then bears a replica of the original surface in positive relief.

The study of fossil invertebrates is important for evolutionary biology, and in recent years has given data valuable for the reinterpretation of the process of evolution. There is now evidence that much evolutionary development has proceeded by rapid periods of change followed by long periods of consolidation and little change (stasis). The rapid changes may have taken place in small populations isolated from the main distributional area, which were able to expand after the extinction of the parent population.

Traditionally, however, it is vertebrate paleontology that has contributed most to evolutionary biology. The transitions from lungfish to amphibians in the late Devonian, from amphibians to reptiles in the late Carboniferous, from mammal-like reptiles to mammals in the Mesozoic, and from a group of diapsid reptiles to birds in the late Jurassic, are some of the classic histories of comparative anatomy and paleontology. So is the diversification of mammals in the Tertiary after the demise of dinosaurs, for the Mesozoic mammals were seldom of any great size, probably because of the dominance of the reptiles at that time. It was only in the Tertiary that the great evolutionary diversification into a number of different ecological roles (adaptive radiations) took place that led eventually to the plethora of present-day mammals.

An equally compelling sequence of events can be seen in the history of the land plants (paleobotany), from small beginnings in the Silurian, through bryophytes (mosses etc), pteridophytes (ferns etc), to gymnosperms and angiosperms (flowering plants).

Micropaleontology deals with very small fossils. Some of these, such as foraminiferids, ostracods, conodonts, spores and pollen, are of very great value in stratigraphy, and also in paleoclimatological and paleoenvironmental studies. There are many micropaleontologists employed by oil companies undertaking essential work on oil-bearing Mesozoic and Tertiary rocks.

Much is now becoming known about the life of the Precambrian, and fossil bacteria and blue-green algae are known as far back as about 3,300 million years. This also largely falls within the domain of the micropaleontologist, and is one of the most thriving fields of paleontology.

Invertebrate fossil animals, like those of today, were finely

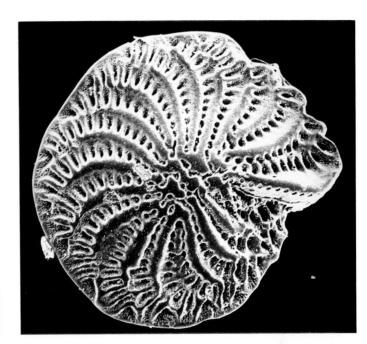

◄▼► **An enormous range** of life forms are preserved as fossils. Ammonites were once one of the most flourishing groups of invertebrates. They became extinct at the end of the Cretaceous period. LEFT *Promicroceras marstonense* from the Lower Jurassic (213–188 million years ago).

Most fossil plants were originally deposited in freshwater lakes or on deltas. BOTTOM LEFT *Sphenopteris obtusiloba* from coal measures of the Upper Carboniferous (320–286 million years ago). Even minute organisms, such as bacteria and protozoans, appear in the fossil record. RIGHT a microfossil magnified 700 times. BELOW a fossil vertebrate; a reptile (*Thaumaturus intermedius*).

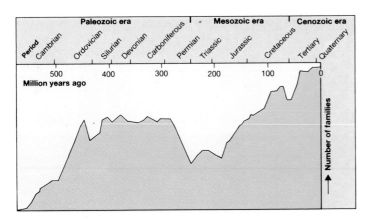

▲ **Prehistoric animal populations.** This graph indicates the relative abundance of animal life in each prehistoric period: it shows the numbers of families of marine organisms. Major extinctions occurred at the end of the Permian and Cretaceous periods. These are used to define the ends of the Paleozoic and Mesozoic eras, times that were important in world history for changing the faunas of the earth.

adjusted to their environments. The relation between fossil and the enclosing rock gives information not only to paleontology but also to the study of sedimentary rock formation, and helps in unraveling the nature of the whole environment (paleo-ecology). In certain cases, therefore, fossils are of great value in interpreting the kind of conditions in which a particular suite of sediments formed (see overleaf). Moreover, fossils are found in different assemblages, which are characteristic of particular environments, and the study of the ancient communities, of which the assemblages are a reflection, and of the food webs and other ecological relationships, falls within the domain of paleoecology.

Likewise paleobiogeography, which deals with the large-scale distribution of animals and plants over the earth's surface in ancient life-provinces, has links with ecology, and also provides data that is useful in interpreting paleoclimates and also, to some extent, the former relative positions of continental masses.

While paleontology remains primarily related to geology, it has also given biological science an essential element: a time perspective against which evolutionary changes may be seen and understood.　　　　ENKC

THE GEOLOGICAL TIME SCALE

One of the chief legacies of 19th-century geology is the geological time scale: the division of geological time into named intervals separated from each other by major changes in rock type, obvious breaks in the succession, and abrupt changes in fossil groups. The coarsest division is the eon, then come eras, periods and epochs.

Archean Eon
The very ancient eon. Between the formation of the earth and 2.5 billion years ago. Some geologists place the oldest limit at the time of the oldest rock, 3.8 billion years, and class everything before this as the Hadean Eon.

Proterozoic Eon
The eon of first life. 2.5 billion to 590 million years ago. It is a time when life is known to have existed but left no clear fossils. The 2.5 billion year boundary has been set by the radiometric dating on a number of igneous and metamorphic events. However, it is now evident that life existed for some time before that.

Phanerozoic Eon
The eon of obvious life. 590 million years ago to the present day. This is the time from which good fossils are known, due to the evolution of hard shells and skeletons at the beginning of the eon.

Precambrian times
Everything before 590 million years ago. Encompasses the Archean and Proterozoic Eons — 80% of the earth's history.

Paleozoic Era
The era of ancient life. Begins with evolution of animals with hard shells and skeletons. Ends with an extinction of most of the marine fauna. Climate mostly warm but with short ice ages. Continents moving together.

Mesozoic Era
The era of middle life. The age of reptiles. Ends with the extinction of the great reptile types and much of the marine fauna. Climate warm throughout. Continents joined together as Pangea, but starting to move apart late in the era.

Cenozoic Era
The era of recent life. The age of mammals and of Man. Climate deteriorating towards recent ice age. Continents continue to move apart.

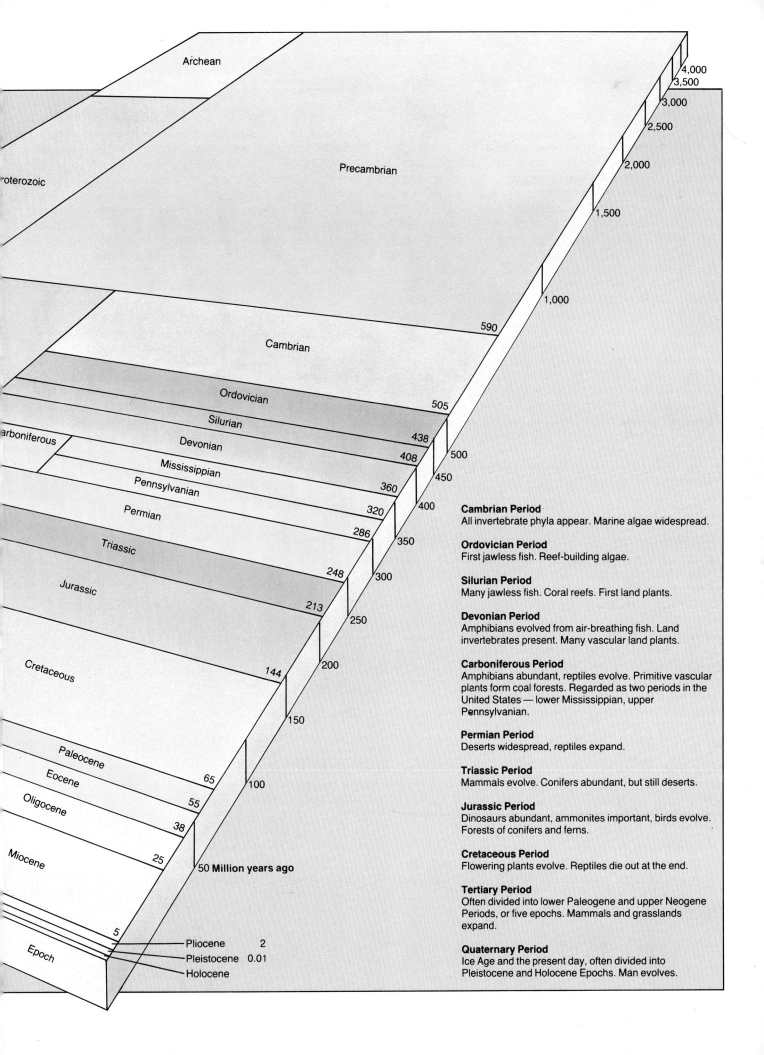

Archean

Precambrian

Proterozoic

4,000
3,500
3,000
2,500
2,000
1,500
1,000

590

Cambrian

Ordovician   505

Silurian   438
Carboniferous
Devonian   408   500
Mississippian   360   450
Pennsylvanian   320   400
Permian   286   350
Triassic   248   300
Jurassic   213   250
                 144   200
Cretaceous   150
Paleocene   65   100
Eocene   55
Oligocene   38
Miocene   25   50 **Million years ago**

5
Pliocene   2
Epoch   Pleistocene   0.01
Holocene

Cambrian Period
All invertebrate phyla appear. Marine algae widespread.

Ordovician Period
First jawless fish. Reef-building algae.

Silurian Period
Many jawless fish. Coral reefs. First land plants.

Devonian Period
Amphibians evolved from air-breathing fish. Land invertebrates present. Many vascular land plants.

Carboniferous Period
Amphibians abundant, reptiles evolve. Primitive vascular plants form coal forests. Regarded as two periods in the United States — lower Mississippian, upper Pennsylvanian.

Permian Period
Deserts widespread, reptiles expand.

Triassic Period
Mammals evolve. Conifers abundant, but still deserts.

Jurassic Period
Dinosaurs abundant, ammonites important, birds evolve. Forests of conifers and ferns.

Cretaceous Period
Flowering plants evolve. Reptiles die out at the end.

Tertiary Period
Often divided into lower Paleogene and upper Neogene Periods, or five epochs. Mammals and grasslands expand.

Quaternary Period
Ice Age and the present day, often divided into Pleistocene and Holocene Epochs. Man evolves.

Footprints from the Past
Trace fossils

Paleontology includes not only the study of "body fossils," that is the actual remains of organisms, but also structures in the enclosing sediment made by the animals when they were alive. These are trace fossils, and their study and that of their modern representatives is known as *ichnology*. Although geologists have studied trace fossils for more than 150 years, it is only comparatively recently that they have been recognized both as sensitive indicators of past environments and as vital clues to the behavior of the animals that made them.

Trace fossils include tracks and trails on the surface, borings in hard substrates, and burrows, mines and galleries in soft ones. They are usually preserved at the interface of two different kinds of sediment. For example, a marine arthropod walking over the surface of firm mud will leave a characteristic trail of leg imprints behind it on the sediment surface, but this will be obscured and will not be preserved if more mud layers accumulate on top. If, on the other hand, silt or sand is deposited shortly after the trail has been formed, then the imprints are filled and permanently preserved by the lower surface of the sand bed as it hardens. When the mud (which is eventually hardened to shale) is washed off the lower surface of such a slab the imprints are visible in negative relief on the bottom of the sandstone.

Trace fossils give much information about the nature of the sedimentary environment in which they were formed, and in some respects they have advantages over other fossils. One such advantage is that they are "in place"; they could not suffer any degree of transportation without being destroyed. This means that the trace fossils must have been part of a fossil community living at a particular time in the place where they are preserved. So often shells and other body fossils have drifted in currents or been washed far away from where they lived, and in such cases are not of great value in interpreting the nature of the environment. Frequently trace fossils occur in sandy or silty rocks where other fossils have been dissolved.

Whereas most trace fossils have a long time range and hence are of limited stratigraphic value, they are often restricted to a very narrow environmental (or facies) range, and thus can be very helpful in interpreting the environment in which they formed. Particular kinds of traces, irrespective of the animals that produced them, occur in certain facies only. Thus in shallow waters many animal species form burrows of various kinds within the sediment so that they can hide from predators; alternatively they can produce shallow resting burrows. The burrows may be vertical tubes, but are more commonly U-shaped with an entrance for food and oxygen-bearing water and an exit for excreta and waste gases or for escape. In somewhat deeper water and in conditions of quieter sedimentation deposit feeders are abundant, and these produce feeding burrows and mines or galleries. In this deeper water where some light penetrates there are a number of surface-crawling traces, but these become particularly abundant in deep water beyond the limit of light penetration where there is no advantage in hiding in the sediment since the animals cannot be seen. In this deep-sea facies are found deposit-feeding and grazing trails, often of complex form. Such trace fossil associations occur in sedimentary rocks ranging back as far as the Cambrian, and an analysis of the "ichnospectrum" helps in understanding their original environments. A facies dominated by surface-grazing trails is more likely to have been of deep-water origin beyond the reach of light. This comparison is borne out by photographs of the abyssal plains of the deep sea, which show very large spiral surface-grazing traces, similar in size and pattern to those of ancient deep-sea deposits going back to the Carboniferous (ie 330 million years ago). If there are no trace fossils at all in a sediment the seafloor may have been stagnant.

▲ **Burrows from the Upper Jurassic** (163–144 million years ago), from Cap Griz Nez, Boullonais, northern France. They were probably made by a decapod crustacean (*Thalassinoides*).

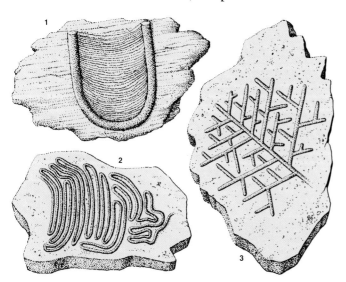

◄ **Some examples of trace fossils.** (1) A U-shaped tube. Its maker lengthened and deepened the tube by removing sediment from the floor of the burrow and plastering it against the ceiling. (2, 3) Systematic searching patterns.

► **Lines interpreted as behavior:** ABOVE fossil trails which have been interpreted BELOW as traces left by mating horseshoe crabs (genus *Limulus*).

▲▼ **The most spectacular trace fossils** are the footprints left by large extinct animals, such as dinosaurs ABOVE. Some fossil footprints are formed as shown BELOW. (1) A footprint is left in soft mud. (2) It is covered with loose sand. (3) The sand is consolidated into sandstone and the mud compacted into shale. (4) When the adjoining beds are split apart the original footprint is revealed in the shale and its cast in the sandstone.

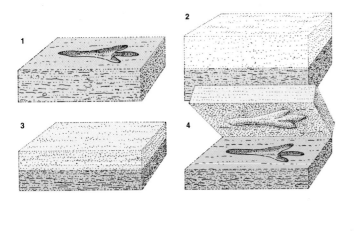

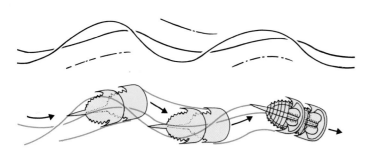

Not only are trace fossils invaluable in sedimentology, they are effectively "fossilized behavior" and are usually the only means available for knowing how ancient animals moved and behaved. The most interesting examples are those made by worms and by trilobites. There are, for instance, many kinds of surface traces made by worm-like animals feeding on an organic-rich surface-layer and adapted to cover the ground as effectively as possible. For example, a spiral track with closely set whorls seems to represent an effective way of mopping up a rich food patch. Another kind of trace consists of closely-spaced meanders in which the animal *Helminthoida* moved some distance along a straight track, turned in a tight loop, and then continued parallel with the first track before turning again; the whole economically effective process was repeated several times until foraging was finished.

A further kind of behavioral system is recorded in nearshore sediments subject to alternate rapid deposition and erosion of sand or silt. In these unstable sediments lived *Diplocraterion*, animals of unknown type inhabiting long U-shaped tubes. These animals liked to keep at a constant distance below the sediment surface. If the sediment was rapidly eroded they moved down, excavating a new lower part of the tube and leaving a series of concentric meniscus-markings (a "spreite") between the parallel arms of the tube. If, on the other hand, sediment was piled on top they moved up, and a spreite was left below the final position of the U-tube. Some examples have moved up and down as the sediment was being deposited or eroded, and this remarkable fossil assemblage testifies not only to the behavior of the animals that made it but also to the nature of the high-energy and unstable environment in which they lived.

ENKC

PRECAMBRIAN AND PALEOZOIC LIFE

How life probably began. . . The earliest living cells: blue-green algae and other bacteria. . . Microfossils and stromatolites: evidence of Precambrian life. . . The evolution of larger-celled and multicellular organisms. . . Ediacaran faunas: the first metazoans. . . The appearance of animals with hard parts. . . Invertebrate diversification and adaptive radiation. . . Coral-reef and sea-bed faunas. . . The earliest fish. . . The exceptional preservation of bacterial and soft-bodied biotas. . . The evolution of life on land: plants, invertebrates, amphibians

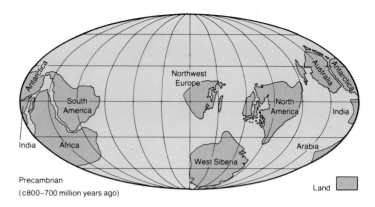

Precambrian
(c800–700 million years ago)

Land

THE contrast between this planet and our lifeless neighbors Mars and Venus is striking; instead of sterile rocky landscapes our oceans, land and air teem with an immense variety of life. Yet the earth was surely not always like this; initially it too must have been devoid of life. The earth formed about 4,600 million years ago, but the earliest stages of its history are shrouded in obscurity. The oldest rocks yet identified occur in west Greenland and are about 3,800 million years old, while the earliest reasonable evidence for life dates from approximately 3,500 million years ago. Life evolved, therefore, within 1,000 million years of the earth's formation.

Details of the early evolution of life remain elusive, but the outlines of the original story are beginning to emerge. Prebiotic reactions (ie chemical reactions that proceed without the agency of living material), perhaps in warm shallow seas and lakes or on land in association with clay minerals, created the basic building blocks of life (such as proteins, lipids and nucleic acids) that ultimately led to the formation of a replicating cell, probably little different from certain modern bacteria. Sources of energy needed to drive these prebiotic chemical reactions may have included volcanic and radioactive heat from the earth, lightning from thunderstorms, and ultraviolet radiation, perhaps augmented by an episode of major meteorite impacts prior to 4,000 million years ago. Traces of these collisions have long since disappeared on earth owing to the action of weathering, erosion and tectonic recycling, but on the airless Moon and other planets the scars remain. In the laboratory heating, electric discharges, ultraviolet radiation and pressure shocks mimic these energy sources, and have been applied to different mixtures of gases and water chosen in the hope that they may approximate to the earth's early atmosphere (see p81). These and other laboratory experiments have replicated the prebiotic synthesis of both simple and relatively complex organic molecules, and have even hinted at the mode of formation of the cell.

There is debate about whether gases such as ammonia (NH_3) and methane (CH_4) were present in the early atmosphere, but there is near unanimity that oxygen was almost absent. The development of an oxidizing atmosphere is due to release of oxygen during photosynthesis, perhaps augmented by a process known as photodissociation by which sunlight cleaves water into its constituent hydrogen and oxygen. How fast oxygen was added to the Precambrian atmosphere is debatable, but many aspects of early evolution are best explained in the context of rising levels of oxygen.

The first living cell was certainly prokaryotic rather than a eukaryote. Prokaryotes today are represented by a wide variety of bacteria and blue-green algae, kingdom Monera. The four eukaryotic kingdoms comprise the protistans (unicellular forms and their immediate multicellular descendants), fungi, plants and animals. The differences between prokaryotic and eukaryotic cells are profound. The former are generally smaller. They lack a well-defined nucleus and other membrane-bound cytoplasmic organelles like mitochondria and chloroplasts, as well as golgi bodies and flagellae. The hereditary material (DNA) is not organized in regular paired chromosomes, and therefore when prokaryotic cells divide there is no well-orchestrated separation of the chromosome pairs (mitosis). Sexual reproduction is unknown among prokaryotes, whereas in eukaryotes the halving of the chromosome complement (meiosis) is later followed by recombination as the nuclei of two gametes fuse during fertilization. Eukaryotes thus have a more complex organization, and they evolved very much later in the Precambrian.

Among the prokaryotes the cyanobacteria (or blue-green algae) were significant in many Precambrian biotas (ie the systems of life then present). Unlike other photosynthetic bacteria the oxygen they generate is liberated as free gas; much of the rise in oxygen levels during the Precambrian was presumably a result of cyanobacterial photosynthesis. Cyanobacteria show tolerance of a wide range of environments, and in particular their resistance to ultraviolet radiation may be a holdover from the early Precambrian when no protective ozone layer screened the earth's surface. Other bacteria also thrived during the Precambrian; modern bacteria occupy practically every known environment, some even flourishing in hot-springs with a pH of 2 (equivalent to boiling sulfuric acid), and similar forms were doubtless well adapted to some of the less clement Precambrian environments.

Much of the evidence for Precambrian life has hinged on the discovery of fossil microorganisms in chert. These microfossils require special conditions for their preservation, but some of them were also responsible for constructing domed or columnar sedimentary structures with characteristic laminations, known as stromatolites. Today stromatolites have a limited distribution, restricted to areas free of grazing and burrowing animals, but before animals evolved, stromatolites were far more widespread and abundant. Most stromatolites are constructed primarily by cyanobacteria, but a variety of other prokaryotes give an exceedingly complex microbial community

◄ **The beginning of the earth.** The present number and arrangement of the earth's continents are the outcome of a long and complex history. The earth is thought to have formed about 4,600 million years ago. In the Precambrian era, that is the next 4,000 million years, the earth developed the geological foundations of the modern world. Its internal structure became ordered into a core, mantle and crust. An atmosphere also developed and rain began to fall, which could form rivers, lakes and seas and erode crustal areas.

The oldest known rocks date from fairly early in the era, about 3,800 million years ago, as do the earth's original continents, which began to form about 3,500 million years ago, though large areas of these remained submerged. A thousand million years later about half of the present continental areas had been created. In the remainder of the era original rocks were deformed and metamorphosed and overlaid by further deposits. The map shows the disposition of the continents late in the Precambrian era, about 800–700 million years ago.

▼ ► **Life, 3,500 million years ago.** Some of the earliest living forms known on earth are fossils of structures called stromatolites RIGHT. They preserve layers of mucilage sheaths produced by cyanobacteria or blue-green algae. Their formation can be studied in living examples, such as the stromatolites found today off the west coast of Australia BELOW. The microorganisms that gave rise to stromatolites made a vital contribution to the growth of life: capable of photosynthesis they emitted oxygen, developing the planet's atmosphere. The oldest known stromatolites occur in rocks dated to 3,500 million years ago.

Fossils in Abundance

The process of bacterial decay and scavenging means that the paleontologist is at best provided with an impoverished sample of the original biota. Occasionally, however, exceptional physiochemical conditions arise that ensure the preservation of delicate organisms. These biotas are of immeasurable importance in evolutionary studies.

Within the Precambrian early silica permineralization, especially of stromatolitic horizons, has preserved a wide variety of cellular remains. Perhaps the two most important are the Gunflint Chert (Canada) and Bitter Springs Chert (Australia). The discovery of the Gunflint microbiota, which is about 2,000 million years old, was serendipitous. An economic geologist studying thin sections of the Chert noticed a profusion of superbly preserved microorganisms. Cyanobacteria are abundant, while other prokaryotic microfossils appear to represent budding manganese bacteria, today known from freshwater lakes. Another Gunflint bacterium also has modern equivalents, some examples turning up in urine-soaked soil beneath a wall of Harlech Castle in Wales! The younger Bitter Springs Chert (900 million years old) is also rich in well-preserved cyanobacteria, but there is disagreement about whether some large unicells represent primitive eukaryotes.

Late Precambrian rocks are famous for the Ediacaran assemblages. Among the most spectacular of Ediacaran faunas are those in southeast Newfoundland where entire seabeds have been smothered intact by volcanic ash falls. Preeminent among younger soft-bodied faunas is the Middle Cambrian Burgess shale of British Columbia. The superbly preserved fauna includes sponges, cnidarians, mollusks, echinoderms, polychaete and priapulid worms, together with an immense variety of arthropods. In addition there are enigmatic animals that represent extinct phyla, and one of the earliest fossil chordates. In Upper Cambrian nodules in Sweden there is another remarkable fauna of minute arthropods, including primitive crustaceans, that have been preserved as three-dimensional replicas in phosphate.

Among the most important Upper Paleozoic faunas are the Hunsrückschiefer (Devonian, West Germany) and Mazon Creek (Carboniferous, Illinois). In the former deposit the fossils have been replaced by iron pyrites and so are readily studied by X-ray radiographs. The style of preservation of the Mazon Creek fauna is very different, the fossils being located within ironstone nodules. The diverse fauna includes many soft-bodied worms and arthropods. The star attraction, however, is a bizarre animal with a grasping feeding appendage and a pair of eyes located on an elongate transverse bar.

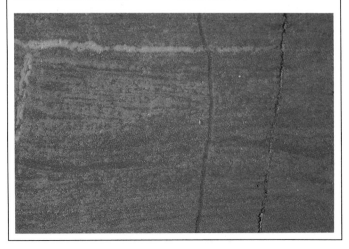

within the stromatolite. On the surface of the stromatolite a mat of filamentous cells of cyanobacteria, encased in sheaths of sticky mucilage, traps sediment grains brought in by tides and currents and so the mat is buried. To resume photosynthesis the cells glide upwards until they lie on top of the newly deposited sediment. Repeated episodes of burial and migration produce the distinctive stromatolitic laminations, organic-rich boundary layers representing mucilage sheaths abandoned as the cells glided upward through the intervening sediment layers.

Fossil microorganisms and stromatolites provide direct information on the early evolution of life, but there is also indirect evidence on Precambrian biological activities. Organic molecules released during decay may survive in the sediment. Moreover, during metabolism organisms often preferentially select between carbon isotopes ^{12}C and ^{13}C. By measuring the ratio of these two isotopes in samples of Precambrian carbon it is sometimes possible to infer the existence of life even though no actual fossils are present.

The oldest generally accepted fossils occur in 3,500 million-year-old rocks in northwest Australia, near a place named North Pole. They are small stromatolites that grew in shallow lagoons, set in a volcanic landscape, which periodically dried out so that gypsum crystallized on the mudflats. A number of other stromatolites older than 3,000 million years are known, but they are rare. While unchanged sediments of this immense age are scarce, most having been caught up in the vicissitudes of the recycling of the earth's crust, the paucity of stromatolites may also reflect a lack of suitable environments. During the early Precambrian the earth's crust seems to have been unstable, with massive volcanic activity. Continents were small, and there were probably no broad continental shelves flooded with shallow sunlit seas. About 2,500 million years ago, however, the earth's crust seems to have stabilized and it is probably no coincidence that stromatolites also increased in diversity and distribution as wide continental shelves formed. Stromatolites continued to flourish for much of the Precambrian, but about 700 million years ago they declined in diversity, perhaps owing to the rise of multicellular animals and their destructive burrowing and grazing.

Sometimes associated with stromatolites are remnants of microorganisms, of which perhaps the most important are those from the 2,000-million-year-old Gunflint Chert. Although the Gunflint Chert provides a landmark in Precambrian paleobiology, other microbiotas in cherts and more rarely other rock types provide further information on prokaryote diversity and evolution. Next in importance to the Gunflint is the microbiota from the Bitter Springs Chert, exposed near Alice Springs in central Australia, consisting of stromatolitic layers containing a wide variety of cyanobacteria.

Although evolutionary changes have been identified during the Precambrian, the overwhelming impression of cyanobacterial evolution is one of stagnation. The simple morphology of cyanobacteria, however, could conceal revolutionary changes in the physiology that left no direct record in the fossilized spheres and filaments. Such apparent conservatism has been labeled the "Volkswagen Syndrome," in analogy to an unchanging exterior design concealing revolutionary internal

◀ **Precambrian Gunflint Chert** from Canada, BELOW bearing traces of blue-green algae.

▼ **From the earliest faunas** of the Precambrian world, a sea-pen-like Ediacaran fossil; from southeast Newfoundland, Canada.

▼ **The acquisition of hard parts** BELOW was a notable development of animal life. External and internal molds of brachiopods; Ordovician period; North Wales.

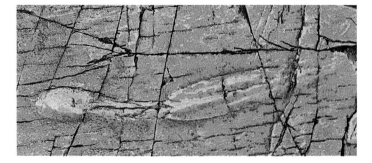

changes. Much of the Bitter Springs biota is prokaryotic, but there are also large cells with dark inclusions that have been interpreted as eukaryotes preserved with organelles. Although the cell size is indicative of eukaryotes, it seems possible that the supposed organelles are nothing more than collapsed protoplasmic residues formed during decay. Laboratory experiments studying the degradation of cyanobacteria can replicate many of these features.

Whatever doubt surrounds the presence of eukaryotes in the Bitter Springs chert, it seems likely that they had evolved at least 1,300 million years ago. For example, rocks of this age in Montana contain carbonaceous films that if identified correctly as seaweeds must represent eukaryotic plants. Even if the fossil record gives a reliable indication of their original time of appearance, it provides no information on how eukaryotes actually evolved from prokaryotes. Comparative studies of modern organisms suggest, however, that the more complex eukaryotes evolved in part from symbiotic associations. For instance, respiratory organelles (mitochondria) may have originally been aerobic bacteria that became incorporated into a larger cell and ultimately lost their independence as free-living organisms.

In general prokaryotes reproduce by binary fission (splitting into two equal parts); sexual reproduction is unknown and exchange of genetic material is unusual. Many eukaryotes can also propagate asexually by simple cell division (mitosis), but most also reproduce sexually. The evolution of sex appears to be one of the keys to eukaryotic success and represents a major landmark in organic evolution. Asexual reproduction produces a clone, a series of genetically identical individuals that inherit all the advantages and disadvantages of the original cell. In eukaryotes, however, the genetic consequences of sexual reproduction are of immense importance.

With the appearance of the eukaryotic cell the stage was set for the evolution of multicellular organisms, notably plants and animals (metazoans). Until the evolution of vascular land plants in the Paleozoic relatively little is known about the earlier evolution of plants as algae. The fossil record of multicellular animals, however, is more complete and effectively begins about 700 million years ago. Evidence on their early evolution comes not only from body fossils, but also from remains of their activities as they moved over or within the seabed preserved as trace fossils (see p10). Many claims for animal fossils older than 700 million years have not withstood scientific scrutiny. For example, structures interpreted as 1,000 million-year-old trace fossils in the Zambian copper belt appear to represent the handiwork of modern termites. Other ancient trace fossils, however, have not been satisfactorily explained and may push the origin of animals further back than 700 million years. Indeed, indirect evidence from molecular biology suggests that animals may have evolved at least 1,000 million years ago.

The first clear insight into early animal history is given by the late Precambrian Ediacaran faunas; these range in age from about 700 to 600 million years old. Some of the first discoveries were made in the Ediacaran Hills of South Australia, but such faunas are now known from most parts of the world. An intriguing similarity between faunas in Australia and northern Russia suggests a cosmopolitan distribution of some Ediacaran species in shallow marine seas. As well as flourishing in shallow waters there were deep-sea faunas, periodically smothered by falls of volcanic ash. Ediacaran faunas share a number of features. Practically without exception the animals are soft-bodied and lack hard skeletal parts. They are dominated by cnidarian-like organisms (relatives of sea anemones and corals). Particularly abundant are various medusoids (jellyfishes), only some of which are clearly related to living groups. Other fossils more or less resemble sea-pens or pennatulaceans, but exact correspondences are difficult to prove. Most problematic are large sac-like organisms, best known from Namibia, which appear to have been embedded in the seabed with a gaping mouth-like aperture. There are, however, other animals that may have been ancestral to higher groups such as arthropods, annelid worms and echinoderms (ancestral to starfishes etc). Although trace fossils are known in some Ediacaran faunas, few if any can be linked to known body fossils, and presumably they represent the activities of yet other soft-bodied animals.

The basic inventory of Ediacaran life is probably fairly complete, but outstanding questions remain. These faunas tell us little about the origins of animals, which conceivably occurred much earlier in the Precambrian. The abundance and diversity of cnidarian and cnidarian-like organisms accord with their generally accepted primitive status. Ediacaran faunas, however, are almost silent on the origins of other animal groups that were to dominate the Paleozoic seas. For example, no ancestors of the mollusks or brachiopods have been recognized. Even among the possible arthropods, none is an obvious ancestor of the highly successful trilobites.

Near the base of the Cambrian (about 590 million years ago) a remarkable event in the history of life occurred: the acquisition of hard parts such as the shells of mollusks and brachiopods, the plate-like skeletons of echinoderms and the carapaces of trilobites and other arthropods. The contrast with Precambrian rocks is dramatic: instead of microscopic remains and scattered stromatolites there is now a rich fossil record. The evolution of hard parts is generally regarded as relatively abrupt, at least in terms of geological time, but this may be in part a result of the way the fossil record was formed. The first animal skeletons may have been secreted as tiny isolated granules and spicules bound by an organic matrix. Upon death these minute structures would be scattered and practically impossible to identify. It was only the later amalgamation of these spicules and granules into coherent skeletal plates and shells that gave them a high chance of fossilization.

The appearance of hard parts is only one aspect of a wider animal diversification that also involved many soft-bodied groups. Although numerous hypotheses have specifically concentrated on why hard parts first evolved, many also have a bearing on wider questions of early animal evolution. Among the most popular are those hypotheses either linked to changing levels of oxygen and carbon dioxide or the appearance of predators. One effect of oxygen levels on animal diversity and hard parts is reflected in marine basins such as the Black Sea where the bottom waters are poisoned with hydrogen sulfide in contrast to the oxygenated surface waters. A transect across the seabed toward the shore reveals dramatic changes in animal diversity. In the deeper water, where oxygen is absent, there are no animals. In slightly shallower water where oxygen concentrations are still very low the only animals are soft-bodied worms, whereas in the oxygenated shallows there is a greater diversity with many animals possessing hard parts. The changes along this seabed transect crudely mimic events in the late Precambrian and early Cambrian with the appearances of soft-bodied animals followed by the evolution of hard parts, suggesting that the amount of oxygen was exerting a controlling influence. While the appearance of animal hard parts is fairly well documented, it is curious that eukaryotic algae and cyanobacteria also developed the ability to secrete calcium carbonate at about the same time. This could be coincidental, but it could be explained by a decrease in concentration of atmospheric carbon dioxide, favoring organic precipitation of calcium carbonate. As hard parts frequently confer protection, it is not surprising that their appearance has also been linked to defense against predators.

The underlying pattern of early animal diversification can be best appreciated by plotting the number of animal types (taxa) against geological time. At the level of family or order

the initial increase in diversity was more or less exponential. Animal diversification did not continue unchecked, and by middle Cambrian times (about 550 million years ago) the curve had flattened out, albeit temporarily. Concealed within the bald exponential curve of early animal diversification is a dramatic set of adaptive radiations during which most and perhaps all animal phyla evolved. The earliest skeletalized faunas in the Lower Cambrian (Tommotian) form a highly distinctive assemblage dominated by a wide variety of small shelly fossils, many of them composed of calcium phosphate. These Tommotian faunas are best known from Siberia, especially along the Aldan

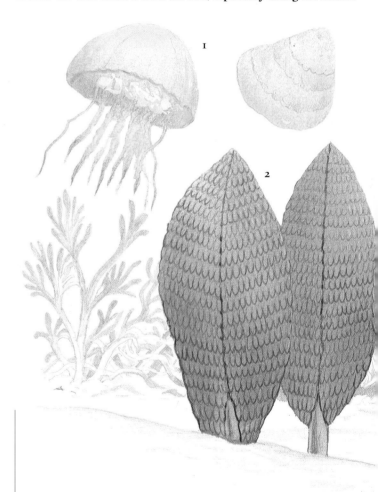

▲ ▶ **A reconstruction of an Ediacaran seabed scene;** late Precambrian era (700–600 million years ago). (1) *Cyclomedusa*, a genus of jellyfish. (2) *Glaessnerina*, a sea-pen-like genus. (3) *Brachina*, a genus of jellyfish. (4) *Dickinsonia*, a worm. (5) *Charnodiscus*, a sea-pen-like genus. (6) A burrowing worm. (7) *Tribrachidium*, an enigmatic genus.

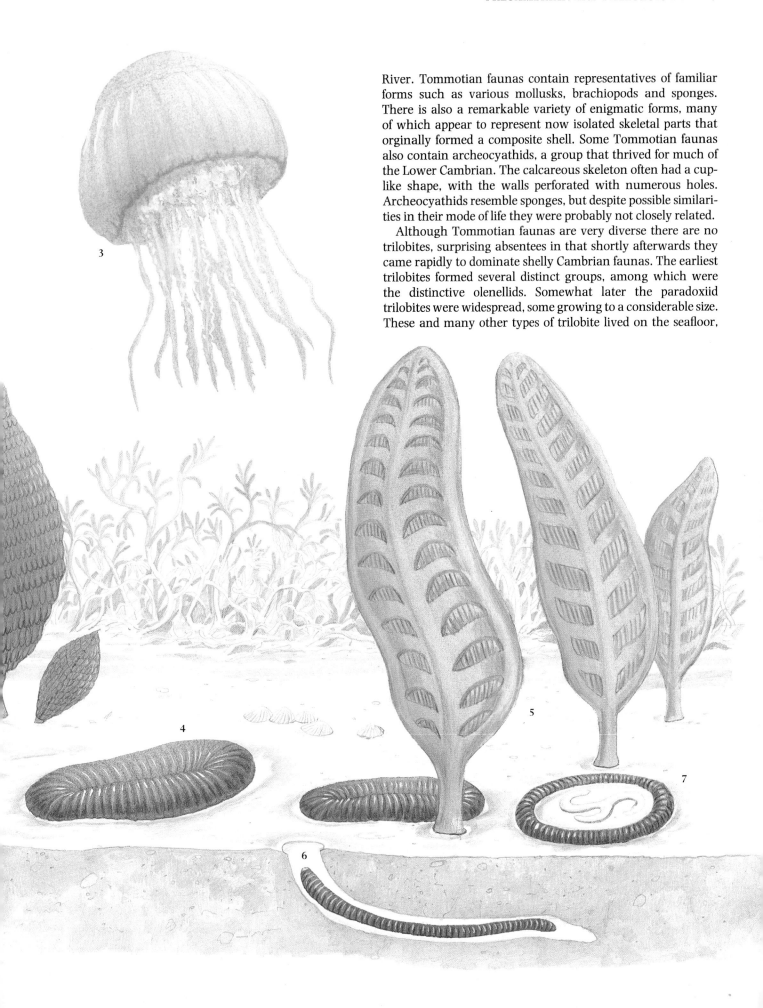

River. Tommotian faunas contain representatives of familiar forms such as various mollusks, brachiopods and sponges. There is also a remarkable variety of enigmatic forms, many of which appear to represent now isolated skeletal parts that orginally formed a composite shell. Some Tommotian faunas also contain archeocyathids, a group that thrived for much of the Lower Cambrian. The calcareous skeleton often had a cup-like shape, with the walls perforated with numerous holes. Archeocyathids resemble sponges, but despite possible similarities in their mode of life they were probably not closely related.

Although Tommotian faunas are very diverse there are no trilobites, surprising absentees in that shortly afterwards they came rapidly to dominate shelly Cambrian faunas. The earliest trilobites formed several distinct groups, among which were the distinctive olenellids. Somewhat later the paradoxiid trilobites were widespread, some growing to a considerable size. These and many other types of trilobite lived on the seafloor,

▶▼ **Reconstruction of fauna** preserved in the Burgess shale of British Columbia, Canada, from the Middle Cambrian period. (**1**) *Pikaia*, a chordate. (**2**) *Eiffelia*, a sponge. (**3**) *Wiwaxia*. (**4**) *Chancelloria*, a sponge-like animal. (**5**) *Vauxia*, a sponge. (**6**) *Scenella*, a mollusk. (**7**) *Dinomischus*. (**8**) Trilobites. (**9**) *Opabinia*. (**10**) *Echmatocrinus*, a crinoid. (**11**) *Aysheaia*, a crinoid. (**12**) *Yohoia*, an arthropod. (**13**) *Pirania*, a sponge. (**14**) *Mackenzia*, a sea-anemone-like animal. (**15**) *Hallucigenia*. (**16**) *Burgess ochaeta*, a polychaete worm. (**17**) *Peronochaeta*, a polychaete worm. (**18**) *Selkirkia*, a priapulid. (**19**) *Ottoia*, a priapulid. (**20**) *Louisella*, a priapulid.

but swimming groups such as the peculiar agnostoids proliferated in the pelagic zone.

The rapid diversification of trilobites is only one aspect of the Cambrian adaptive radiation, and many other groups show comparable features, among which the echinoderm and mollusk radiations are well documented. Cambrian echinoderms show an enormous variety of forms. In some, such as the edrioasteroids, the now diagnostic five-fold symmetry occurs, but other groups range from bilateral symmetry to asymmetry, with one group (helicoplacoids) even having an extraordinary spiral shape. A number of these early echinoderms were confined to the Cambrian, and it is tempting to interpret the profusion of forms as "experiments" in echinoderm design, only some of which were successful. Similar divergences from the original body-plan have been documented in other well-skeletalized groups. For example, in the mollusks bivalves, gastropods and a number of extinct groups diversified from their monoplacophoran ancestors during the Cambrian period. The rare preservation of soft-bodied forms, as in the Burgess Shale, also reveals evidence for tremendous evolutionary diversification. The extent of the Cambrian adaptive radiations is also reflected in the trace fossils, with a dramatic upsurge in types during the early Cambrian reflecting increasing diversity of both species and ecological types.

The lull in metazoan diversification during the middle Cambrian was geologically short-lived, and during the late Cambrian and Ordovician periods marine faunas witnessed a further set of adaptive radiations. Indeed, in terms of the number of families that evolved this diversification dwarfed the early Cambrian explosion. In one sense, however, the evolutionary events were less dramatic because with few exceptions (eg bryozoans, which first definitely occur in the Ordovician) no new phyla evolved; rather the adaptive radiations exploited a

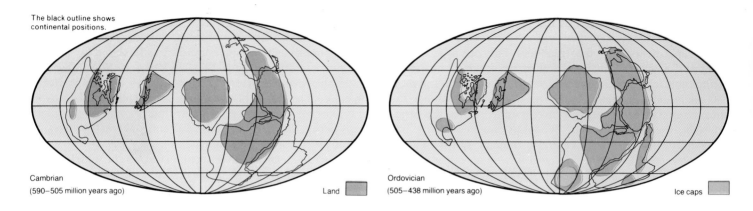

The black outline shows
continental positions.

Cambrian
(590–505 million years ago)

Land

Ordovician
(505–438 million years ago)

Ice caps

Paleozoic Terrestrial Life

From the plethora of marine groups that flourished in the early Paleozoic seas only a handful managed to adapt to life on land, a tribute to how hard it was to do so. Plants apparently preceded animals onto land, with the earliest reliable records of vascular land plants (psilophytes) occurring in the Silurian (438–408 million years ago). By the late Devonian (about 370 million years ago) a variety of distinct groups had evolved from the more primitive psilopsids, including plants with sufficient secondary wood to be called trees. It seems likely that forests had already appeared by this time. The rapid evolution of early land plants saw many additional innovations, including the development of leaves and branching habits, but perhaps most important were those advances connected with reproduction, especially the evolution of seed that permitted the spread of plants into drier habitats.

A unique source of information on early terrestrial life comes from the Rhynie Chert, a remarkable permineralized peat-bog that contains superbly preserved psilopsid plants and a variety of arthropods including primitive insects and arachnids. There is intriguing evidence of animal–plant interaction in the Rhynie Chert, including stems apparently punctured to obtain sap. While plants, arthropods and probably worms flourished and coadapted during the Devonian, the evolution of amphibians from fish appears to have occurred only at the very end of the Devonian. Whether the path of evolution of amphibians was via freshwater rivers and lakes or more directly from the sea is uncertain. Early amphibians must have been closely tied to water, and although later examples show evidence of greater independence, it was the evolution of reptiles with their amniotic eggs in the Carboniferous that enabled the conquest of terrestrial environments to proceed apace. Early amphibians and reptiles appear to be almost entirely carnivorous, probably hunting fish, insects and—when the opportunity arose—each other. By Permian times, however, herbivorous forms had arisen.

The extensive and luxuriant Carboniferous floras are well documented, largely thanks to the immense economic importance of their end-product—coal. Recent studies have shown that the original coal swamps were not one monotonous morass, and that although there were enormous waterlogged swamps with extensive stands of gigantic lycopods, other areas of alluvial floodplain supported diverse floras rich in ferns and seed-ferns (pteridosperms). Within the coal forests amphibians and reptiles lurked, while many types of arthropods flourished. Some of the latter were enormous: *Arthropleura* (3) was a gigantic millipede-like animal that grew to almost 2m (6.5ft) in length; it appears to have been an innocuous herbivore. Winged insects appear to have evolved in the Carboniferous; among these were giant dragonflies (1) as well as more mundane forms such as cockroaches (2). New studies of the fossil record are showing how plants and animals interacted in the Carboniferous forests, and over the next few years much will be learned about these complex ecosystems.

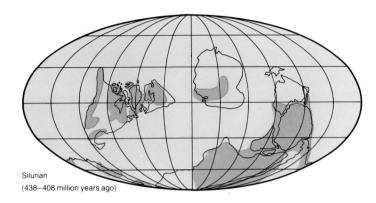

Silurian
(438–408 million years ago)

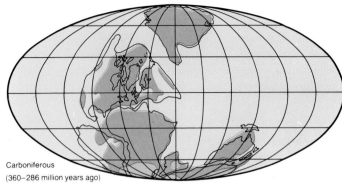

Carboniferous
(360–286 million years ago)

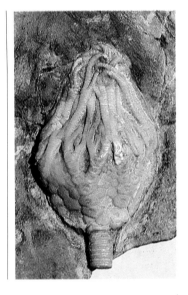

▶ **Successful animals of the Paleozoic** included the crinoids (order Crinoidea), which were at their height in the Carboniferous period (360–286 million years ago). They consisted of root-like outgrowths, which anchored them to the seabed or elsewhere; a stem; and a calyx or cup-like structure, seen here. This enclosed the body cavity. Crinoids achieved a wide range of sizes, from minute species to ones with stems 20m (66ft) long.

◀ **Reconstruction of a forest scene** of the Carboniferous period (360–286 million years ago), containing club mosses, calamite horsetails, cordaits, tree ferns. (1) Giant dragonflies. (2) A giant cockroach. (3) *Arthropleura*, a giant millipede-like animal.

series of already established body plans. Among the groups that flourished were the mollusks, especially bivalves and cephalopods, a wide variety of echinoderms, tabulate and rugose corals, articulate brachiopods and bryozoans. The success of these groups was to set the seal on the character of marine life for the remainder of the Paleozoic, with the type of faunas that had dominated earlier in the Cambrian entering a slow decline in relative importance. This decline has recently been correlated with a progressive offshore migration into deeper waters. Following the Ordovician adaptive radiations, levels of diversity appear to have remained approximately even until the late Permian, when an especially severe mass extinction fundamentally remodeled the character of marine biotas.

With the rise of corals and bryozoans as well as calcareous algae the opportunities for the construction and binding of an organic framework or reef arose. Earlier archeocyathid reefs had briefly flourished, but the later Paleozoic reefs were far more diverse and complex, and provided a variety of habitats that encouraged the further diversification of groups such as

the crinoids and brachiopods. By Silurian and Devonian times immense reef complexes dotted the shallow waters of the equatorial-shelf seas. Reefs continued to be significant during the Carboniferous, while the immense Capitan reef of Permian age is superbly exposed in the Guadaloupe Mountains of New Mexico. In this reef calcareous algae and bryozoans are especially important, but the rich fauna also contains extraordinary aberrant brachiopods that superficially resemble cup-corals. Elsewhere in Paleozoic seabeds areas of carbonate sediment lithified and supported distinctive faunas of encrusting organisms such as crinoids and bryozoans.

While reefs represented the acme of biotic complexity in Paleozoic marine faunas, the enormous areas of seafloor covered by soft sediment supported diverse faunas. Of the soft-bodied organisms exceptional faunas such as those in the Devonian Hunsrückschiefer and Carboniferous Mazon Creek allow unique glimpses into the original richness of life. Many seabeds also supported a flourishing fauna of brachiopods. Various associations of species have been recognized, their distribution apparently being controlled by factors such as sediment type and water depth. These seabed communities often supported flourishing assemblages of both solitary and colonial corals, crinoids, blastoids, and among the mollusks, bivalves, which were especially important in shallow-water environments. During the Ordovician and Silurian (505–408 million years ago) trilobites continued to be important, but in the later Paleozoic they entered a decline in diversity, although other arthropods continued to flourish with the evolution of crustaceans and eurypterids, the latter often armed with ferocious grasping limbs.

Within the pelagic realm, Paleozoic seas saw a series of spectacular changes. Pelagic trilobites included forms with enormous eyes, while the early Paleozoic plankton included myriads of graptolites, colonial organisms that were distantly related to chordates. By the early Devonian (about 400 million years ago), however, pelagic graptolites were extinct, although the bottom-dwelling forms persisted until the Carboniferous. Although the earliest fish are from the Upper Cambrian, relatively little is known about Lower Paleozoic fish. However, diversification during the Devonian and Carboniferous saw the appearance of a wide variety of forms, some heavily armored, and other fast-swimming predators. Of greatest interest, however, are the groups known as the dipnoans and crossopterygians; from the latter the terrestrial tetrapods emerged in the late Devonian. SCM

EARLY MESOZOIC LIFE

The age of reptiles. . . Mammal-like reptiles. . . Crocodiles and early dinosaurs. . . Amphibians. . . Marine animals: plesiosaurs, turtles, sharks, bony fish. . . Geography, climate and vegetation. . . Invertebrates: mollusks, crustaceans, insects, corals, sponges, sea urchins

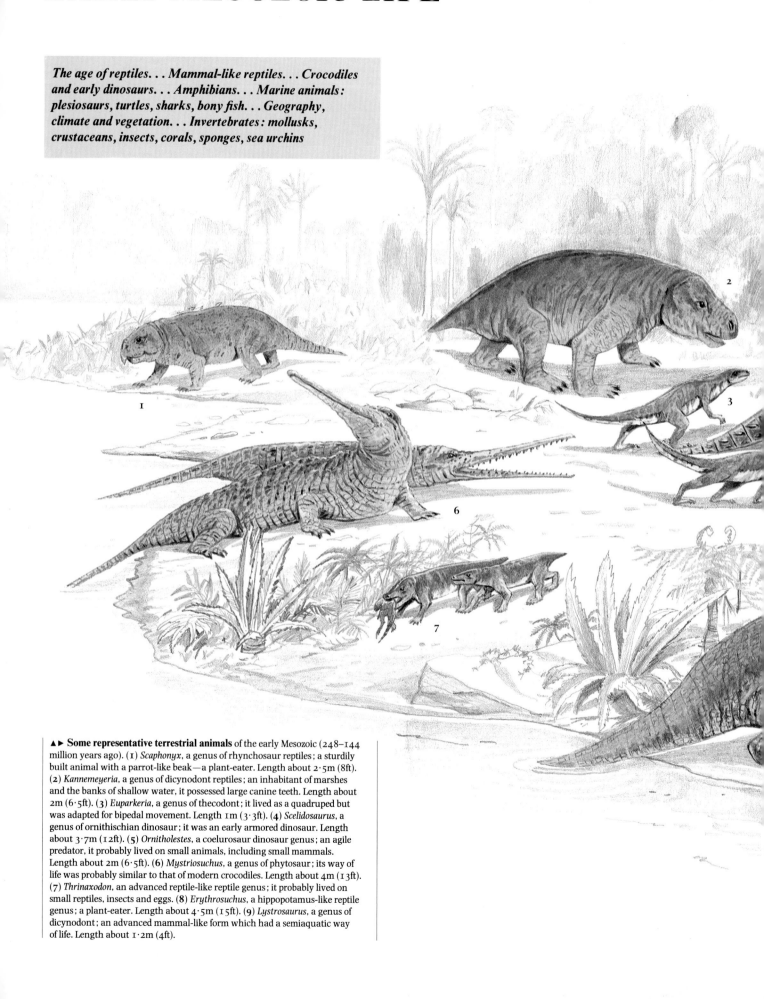

▲▶ **Some representative terrestrial animals** of the early Mesozoic (248–144 million years ago). (1) *Scaphonyx*, a genus of rhynchosaur reptiles; a sturdily built animal with a parrot-like beak—a plant-eater. Length about 2·5m (8ft). (2) *Kannemeyeria*, a genus of dicynodont reptiles; an inhabitant of marshes and the banks of shallow water, it possessed large canine teeth. Length about 2m (6·5ft). (3) *Euparkeria*, a genus of thecodont; it lived as a quadruped but was adapted for bipedal movement. Length 1m (3·3ft). (4) *Scelidosaurus*, a genus of ornithischian dinosaur; it was an early armored dinosaur. Length about 3·7m (12ft). (5) *Ornitholestes*, a coelurosaur dinosaur genus; an agile predator, it probably lived on small animals, including small mammals. Length about 2m (6·5ft). (6) *Mystriosuchus*, a genus of phytosaur; its way of life was probably similar to that of modern crocodiles. Length about 4m (13ft). (7) *Thrinaxodon*, an advanced reptile-like reptile genus; it probably lived on small reptiles, insects and eggs. (8) *Erythrosuchus*, a hippopotamus-like reptile genus; a plant-eater. Length about 4·5m (15ft). (9) *Lystrosaurus*, a genus of dicynodont; an advanced mammal-like form which had a semiaquatic way of life. Length about 1·2m (4ft).

THE Mesozoic era lasted some 183 million years, from about 248 to 65 million years ago. The earlier part of the era comprised the Triassic period (duration some 35 million years, from about 248 to 213 million years ago) and the Jurassic period (duration some 69 million years, from about 213 to 144 million years ago). Amphibians were sparse and of small size, except for the huge Triassic labyrinthodonts and their few Jurassic survivors. Land reptiles are known from abundant faunas of the Triassic age on several continents, those from the South African Karroo being especially impressive, but throughout most of the Jurassic the reptile record is generally poor and improves again only in the Upper Jurassic. Mammals first appeared at the very beginning of the Jurassic, remaining very small throughout the period, while the first known bird (*Archaeopteryx*) made its debut in the uppermost Jurassic of Bavaria. Meanwhile marine deposits became more widespread during Triassic and Jurassic times and provide a more continuous and comprehensive fossil record, mainly of invertebrates.

Physical and Floral Background

The early Mesozoic world was strikingly different from that of the present in two important respects. Instead of there being several continents isolated by ocean there was only one supercontinent: Pangaea, with a northern (Laurasia) and southern (Gondwana) component separated in the east by a tropical seaway, the Tethys Sea. Secondly, the climate was equable. The polar regions lacked ice caps and experienced temperate conditions, as indicated by fossil floras found in Greenland and Antarctica. Conditions comparable to those in the present tropics extended into latitudes as far as 60° from the equator. This is shown, for instance, by the presence of reef corals in marine deposits.

One of the most important consequences of this geography was that most terrestrial animals were far more cosmopolitan in distribution than their living relatives because the physical and climatic barriers to their migration were relatively modest. Thus it is impossible to distinguish faunal provinces for the rich Triassic reptile faunas, whereas today each continent has many endemic vertebrates. Similarly the animals that lived in the shallow seas beside and partly enveloping the

continents were able to migrate far and wide.

Not until well into the Jurassic period, about 150 million years ago, did the supercontinent begin to disintegrate, when a narrow ocean became established between Africa and North America as a result of plate tectonics. This ocean progressively widened through the remainder of the period, but not until Cretaceous times did it extend north and south to separate the Old and New Worlds, when the Indian Ocean also came into existence. The most important geographic changes in the early Mesozoic were in fact involved with changes of sea level. This remained relatively low during most of the Triassic and early Jurassic, but a marked rise of sea level early in the late Jurassic led to a quarter of the present continental area being inundated. This must have served to restrict the free migration of land animals, but nevertheless dinosaur interchange between the northern and southern hemispheres and the Old and New Worlds evidently continued because the late Jurassic faunas of the western United States, Argentina and East Africa show strong similarities.

Today's dominant terrestrial

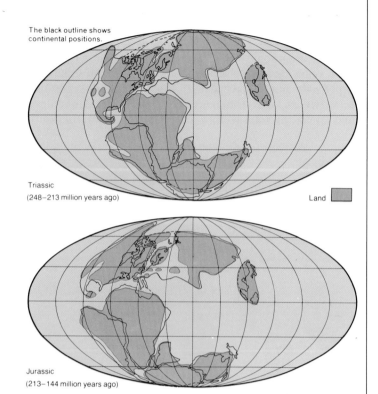

The black outline shows continental positions.

Triassic
(248–213 million years ago)

Land

Jurassic
(213–144 million years ago)

plants, the flowering plants (angiosperms), had not yet evolved. Since they include the great majority of trees and shrubs and all the grasses, the early Mesozoic landscapes had a very different aspect. The most important trees belonged to the so-called gymnosperms ("naked-seed" plants), including conifers, cycad-like

forms and relatives of the maidenhair tree (Ginkgo), whose natural habitat today is confined to China. Ferns were also extremely important, with tropical-type tree ferns extending into high latitudes. Low swampy ground was covered by horsetails, while various mosses and algae were important in damp habitats.

The dominant land vertebrates were the reptiles, the composition of their faunas changing profoundly during Triassic time. Through the first two-thirds of that period the most important reptiles were advanced members of the so-called mammal-like reptiles, the therapsids: the dicynodonts were all herbivores and predominated numerically in the early Triassic, while the cynodonts, which included also carnivores, expanded in importance from the early to the middle Triassic. Both groups faded out in the late Triassic. There can be little doubt that the cynodonts include the direct ancestor of the mammals, though no one really knows which family is involved. Throughout the Jurassic, the mammals (known mainly from their jaws and teeth, the latter being especially resistant to destruction) remained small creatures, no larger than a rat; they might well have adopted nocturnal habits, which ectothermic (cold-blooded) reptiles cannot do, and they would thus have avoided competition with the dominant land vertebrates. By early in the late Triassic the therapsids had been largely replaced by archosaurs, which arose in the early Triassic and

became progressively more important. This major reptile group underwent an evolutionary explosion in late Triassic times; the stem-group (thecodontians) radiated into several daughter groups including dinosaurs of various sorts, crocodilians and pterosaurs (flying reptiles). By late Jurassic times the dinosaurs had reached their maximum size in the huge sauropods, including *Brachiosaurus* (the largest of all) and the well-known *Diplodocus*. At the same time *Megalosaurus* was the dominant large carnivore in Europe, with *Allosaurus* as its North American analog.

Less conspicuous in early Mesozoic land faunas, but doubtless very abundant and therefore ecologically important, were the lizards and their relatives (lepidosaurs). One group of lepidosaurs that enjoyed a brief heyday in the middle and the earlier part of the late Triassic were the rhynchosaurs, large herbivores with beak-like jaws covered in large numbers of tiny rasping teeth.

The phytosaurs were highly predaceous inhabitants of late Triassic lakes and streams, some attaining gigantic dimensions.

They were similar in appearance (and presumably in habits) to modern crocodiles; the main difference was that the phytosaur's nostrils lay in a crater on top of its head, between the eyes, whereas in a crocodile the nostrils are situated at the end of the snout. All the phytosaurs became extinct at the end of the Triassic and were replaced by the true crocodilians, but they could not have been ancestral to the latter. The first thoroughly aerial vertebrates, the pterosaurs, also arose in the late Triassic, but they became really important only in the Jurassic. As for the birds, the first genus known (*Archaeopteryx*) did not make its debut until the late Jurassic, and it was not until well into the following Cretaceous that birds began to play a significant role in the fauna.

A group of ancient amphibians, the stereospondyl labyrinthodonts, were perhaps the most common and widely distributed of all Triassic continental vertebrates; they were highly specialized for an aquatic existence and often very large. As was the case with many other important groups of vertebrates, only a few of them survived the end of the Triassic. The earliest known occurrence of the frog and toad group was in late Triassic times, but their fossil record, by contrast, is very poor.

The early Mesozoic seas also contained an abundance of reptiles—reptiles that had returned to the aquatic medium of their fishy and amphibian ancestors; the most familiar of these are the fish- and squid-eating ichthyosaurs and plesiosaurs. The ichthyosaurs had acquired adaptations comparable to those of whales and dolphins, with a streamlined body shaped like that of a fish, and they were presumably incapable of emerging from the water; their earliest representatives are of early Triassic age. The plesiosaurs, by contrast, were less extremely adapted and could probably come out onto dry land; they did not appear until the end of the Triassic, but they were preceded in Triassic times by the similar but smaller (and presumably ancestral) nothosaurs. Both the ichthyosaurs and the plesiosaurs became abundant in the Jurassic and some members of both groups became very large—even gigantic. Less familiar are the placodonts, Triassic inhabitants of shallow waters, massively built and with teeth adapted for crushing mollusk shells. Turtles are another group that had its origins in the Triassic and expanded in importance in the Jurassic, during which period the crocodilians too invaded the seas.

The abundant fish faunas may be broadly grouped into cartilaginous and bony fishes. Among the former category, the actively swimming predaceous sharks of the Triassic were biologically advanced compared with their Paleozoic predecessors, with newly evolved methods of jaw suspension allowing for increased mobility in the development and function of the all-important jaws. The other cartilaginous fish "evolutionary line," toward flat, bottom-living, mollusk-eating skates and rays, had become differentiated by Triassic times.

Unlike the cartilaginous fishes, the more numerous and diverse bony fishes also colonized fresh water, but the great majority lived in the sea. The primitive chondrosteans, the dominant element in the Triassic, declined severely at the end of the period to be replaced by the holosteans, which in turn gave way to the even more advanced teleosts (the dominant bony fishes today) in the Cretaceous. The lobe-finned fishes, crossopterygians and dipnoans, persisted as a comparatively stable but subordinate element throughout the Mesozoic era.

Although vertebrates tend to attract more attention among the lay public, the collector hunting for fossils in Triassic and Jurassic strata is likely to have far more success in finding the remains of invertebrates because these are far more numerous, especially in marine deposits.

Following the mass extinction of marine invertebrate faunas at the end of the Paleozoic there was a gradual radiation of the survivors so that by mid-Jurassic times family diversity was again as high as in the early Permian. The composition of the faunas had, however, been drastically altered as a result of the extinction and there is consequently no difficulty in distinguishing Paleozoic and Mesozoic faunas. Perhaps the most notable change concerns the brachiopods. This primitive phylum had been the dominant component of the benthos (organisms living on or near the bottom of the sea) through the Paleozoic era, but was one of the groups most severely affected by the mass extinction. Although brachiopods remained one of the commoner invertebrate elements of early Mesozoic faunas they gave way to the bivalve mollusks as the dominant component of the macroscopic benthos.

By far the most spectacular of the shelled mollusks, however, were the ammonites, another group that had been decimated at the end of the Paleozoic. From the comparatively few survivors there was an impressive radiation until in late Triassic times the seas were filled with a varied array of forms, some of them highly ornamented. Then at the end of the Triassic there was another crisis, associated with a major regression of the sea, as in the late Permian, and only one ammonite family survived, to radiate once again into an even more diverse assemblage in the Jurassic. Since the ammonites became extinct at the end of the Mesozoic there is some uncertainty about their mode of life, but they are widely believed, by analogy with the living *Nautilus*, to have been able to swim actively and control their buoyancy within the water column, leading an active predatory or scavenging mode of life.

Although arthropods are probably the most familiar (and certainly the most diverse) of present-day invertebrates, the early Mesozoic fossil record is not a rich one. By far the most abundant were the minute forms with a bivalved carapace, the ostracods, mostly marine but with brackish- and freshwater representatives, and the conchostracans, which flourished mainly in continental waters. The major decapod order, including shrimps, crabs and lobsters, began in the Triassic, as did the unrelated xiphosures or king crabs belonging to the genus *Limulus*, which survives today as a "living fossil." Insects must have been abundant, but we are dependent on unusually good conditions of preservation to find them. The best example is the celebrated Upper Jurassic Solnhofen limestone of Bavaria, which has yielded superbly preserved invertebrates and vertebrates, including the earliest bird, *Archaeopteryx*.

The bryozoans and corals were two other major invertebrate groups that were severely affected by the end-Paleozoic mass extinction, and those elements that occur in the Mesozoic bear no close taxonomic relationship to their Paleozoic precursors. It is likely that the scleractinian corals, which arose in the Triassic and still exist today, originated from some soft-bodied coelenterate ancestor and took over the coral ecological niche

by developing a hard calcareous skeleton after the extinction of the Paleozoic corals. Together with calcareous and siliceous sponges they were the major components of organic buildups or reefs in limestone and dolomite strata. These reefs grew in or adjacent to the tropical belt of the Tethys seaway and remnants of these spectacular constructions can be seen, for instance, in the Triassic rocks of the Alps and the Jurassic rocks of North Africa, western Europe and Poland.

Of the various groups of echinoderms that lived in the early Mesozoic, only the crinoids and echinoids have left a good fossil record. The crinoids, like the corals and bryozoans, belong to new groups which evolved after the extinction of the rich Paleozoic crinoid faunas. They were all attached stalked forms, and flourished in shallow waters, unlike today, when stalked crinoids are restricted to deeper waters. The more mobile sea urchins were initially all regular surface dwellers on the seabed, with spherical outlines, but through the course of the Jurassic the newly evolved ovoid burrowing forms became progressively more important. (A comparable evolutionary change took place at the same time among the bivalve mollusks, with an increasing proportion in the faunas of burrowing siphonate forms.)

Finally mention must be made of two important groups of protistan microfossils, the calcareous foraminiferans and siliceous radiolarians, the latter occurring in huge quantities in the siliceous deposits known as chert, which were laid down in comparatively deep tropical water. The rich record of tracks and burrows known as trace fossils is a reminder of the rich fauna of soft-bodied organisms that had no parts preservable as fossils.

AJC/AH

► ▼ **Aquatic life of the early Mesozoic era.** (**1**) Fish-eating pterodactyls. (**2**) *Peloneustes*, a short-neck plesiosaur. Length about 3m (10ft). (**3**) *Muraenosaurus*, a long-necked plesiosaur. Length 6·5m (21ft). (**4**) *Mixosaurus*, an early ichthyosaur. Length about 2m (6.5ft). (**5**) *Saurichthys*, a fish with a "lurking predator" body form. Length about 18cm (7in). (**6**) *Eurhinosaurus*, a genus of ichthyosaur. (**7**) *Nothosaurus*, a fish-eating reptile, probably a contemporary of the first ichthyosaurs. Length about 3m (10ft). (**8**) *Dapedium*, a deep-bodied fish, covered with heavy scales. Length about 36cm (14in). (**9**) *Placodus*, a genus of placodont reptile with forward-pointing teeth, an adaptation for eating mollusks. Length about 2m (6·5ft).

3

5

6

9

LATE MESOZOIC LIFE

Continued diversification of dinosaurs. . . Evidence of dinosaur biology. . . Pterosaurs and birds. . . Lizards and snakes. . . Primitive mammals. . . Plankton, ammonites and marine reptiles. . . Fish: the rise of the teleosts. . . The diversification of marine invertebrates. . . Physical changes and the appearance of flowering plants. . . Why the mass extinction at the end of the Cretaceous period?

THE late Mesozoic era spanned some 79 million years, comprising but one period, the Cretaceous, which came to a well-marked close 65 million years ago.

Land and Air

On land the reign of the dinosaurs continued, although continental separation and the spread of shallow seas geographically partitioned their later evolution. New ornithischian groups flourished in the Northern Hemisphere. These included the rhinoceros-like ceratopsians and the duck-billed hadrosaurs, both of which reached their acme near the end of the Cretaceous. Among the late Cretaceous predators the fearsome *Tyrannosaurus* was restricted to North America, although a close relative, *Tarbosaurus*, lived in eastern Asia. Meanwhile the descendants of the groups that had flourished in the Jurassic inherited the virtually worldwide distribution of their ancestors. For example, the remaining giant sauropods were widespread in the southern as well as the northern continents. Other old stocks dwindled away: it is doubtful if the stegosaurs survived into late Cretaceous times. Their ecological role seems to have been taken on by the heavily armored ankylosaurs.

Tantalizing glimpses of dinosaur biology come from their fossils. Several features point to social behavior such as parental care of offspring, competition for mates, and herding, particularly among the advanced ornithischians. One of the most exciting clues came from the discovery in 1978 of a fossilized "nest" of juvenile hadrosaurs in some Upper Cretaceous muddy terrestrial sediments in Montana, USA. Eleven little skeletons up to 1m (3.3ft) long were huddled together in an oval depression about 2m across and 75cm deep (6.5ft × 2.5ft), with a further four lying nearby. They were midway between hatchlings and adults in size, and their bone development and the

► **Representative aerial and terrestrial animals** (except mammals) of the late Mesozoic era (Cretaceous period, 144–65 million years ago).
(**1**) *Lambeosaurus*. a genus of hadrosaur or duck-billed dinosaur. Upper Cretaceous of North America. Maximum length including tail about 9m (30ft). (**2**) *Pterodactylus*, a small pterodactyloid flying reptile; it was about the size of a thrush. (**3**) *Alamosaurus*, a sauropodomorph or reptile-footed dinosaur. Cretaceous of North America. Length including neck and tail about 20m (65ft). (**4**) *Triceratops*, the largest genus of ceratopiians or horned dinosaurs. Late Cretaceous of North America. Length about 7m (23ft), weight 9 tonnes. (**5**) Head of a *Tyrannosaurus*, a genus of the carnivorous therapod dinosaurs. Late Cretaceous of North America. Length about 12m (39ft), weight about 7 tonnes; length of skull 1·2m (3·9ft). (**6**) *Ornithomimus*, an ostrich-like therapod dinosaur. Cretaceous of North America. Length about 4m (13ft). (**7**) *Dromaeosaurus*, the smallest genus in the dromaeosaurid family of dinosaurs. Length about 2m (6·5ft). (**8**) *Ankylosaurus*, an armored ankylosaurian dinosaur. Upper Cretaceous of North America. Length 4·5m (15ft). (**9**) *Pachycephalosaurus*, a genus belonging to a group of dinosaurs that had thick skull roofs. Upper Cretaceous of North America. Length 9m (30ft).

wear on their teeth shows that they were already capable of scampering around and foraging for food, so their confinement to a nest strongly suggests parental supervision. Their charmingly appropriate scientific name, *Maiasaura*, means "good mother reptile" in Greek.

Hadrosaur skulls also yield insights on their family life. The curious hollow crests on the skulls of many advanced species have attracted several explanations, but the most likely is that they were resonators for vocal signals. Research on those in *Lambeosaurus* shows that the adults developed internal processes that would have dampened higher harmonics. So their skulls seem to have been adapted so as to give maximum distinction between the bellowing of adults and the piping of juveniles. The ear cavities in the skulls show that the animals would have been able to hear the full range of frequencies that might have been produced from the resonators. Vocal communication between adults and juveniles, like that in living crocodiles, was therefore likely.

Another ornithischian, *Pachycephalosaurus*, had a thick bony skull seemingly adapted for head-butting, thus inviting comparison with some living herd-forming animals who compete for dominance in such matters as mate choice by butting their heads together. More evidence for herding in several kinds of herbivorous dinosaurs comes from multiple parallel tracks of fossilized footprints.

Nowadays there is much debate about the body temperatures of dinosaurs. Could they accurately regulate their temperature, using internal heat, like mammals and birds, to allow activity by day and night in any conditions? Or did their temperature and activity fluctuate with external conditions, necessitating compensatory behavior such as basking in the sun to warm up, as in living reptiles? Or were they somewhere between these two possibilities? Whole books have been written on this difficult subject and there is not space here to do justice to the many arguments put forward: suffice it to say that the issue is far from settled. What does seem to be emerging is that the answer may differ from group to group.

Dinosaur evolution was thus a dynamic story of continuous turnover of diversely adapted species, right up to the close of the Cretaceous. The popular old idea of dinosaurs as lumbering and inefficient brutes destined for extinction is wholly inaccurate.

In the air the pterosaurs continued to thrive. Recent study of the shapes of their forelimb joints shows that they actively flew by flapping their wings, rather than merely gliding, as older reconstructions tended to assume. Their wing membranes were reinforced by tough fibers to stop them billowing, and the hind limbs would have allowed them to hop around on the ground. So they must have been closely analogous with birds, although they were only distantly related. Like albatrosses, some pterosaurs took to soaring, the most famous example being *Pteranodon*, with a wingspan of up to 5m (16ft). Though they lacked feathers, rare fossils show that they had a hairy coating. Such evidence for insulation, combined with that for active flight, suggests close temperature regulation, again as in birds. The birds themselves arose in the late Jurassic, long after the pterosaurs, and possibly from a dinosaur ancestor.

Thriving alongside these in less spectacular fashion were the lepidosaurs, comprising the lizards and their Cretaceous offshoot, the snakes. These, especially the latter, were beginning a phase of expansion that was to outlast the dinosaurs, and indeed continues apace today.

Also playing a subsidiary role were the mammals. Scurrying around, mostly in lowland wood and swamp habitats, though also in some upland woods, they remained rarely more than rat-sized. They were probably mainly nocturnal, and the evolutionary expansion of their brains may owe much to the improved senses of smell and hearing that they evolved in consequence. Their diets mainly concentrated on small invertebrates such as worms, slugs, spiders and insects as well as fruits and seeds. The scrappy but widespread record of these mammals shows that the placentals and the marsupials evolved some time in the early Cretaceous, and lived alongside other, more primitive forms. They diversified much in the latest Cretaceous; and one rich assemblage from Montana, USA, known as the "Bug Creek fauna," is especially fascinating in that it contains several species foreshadowing those of the succeeding Paleocene epoch.

Sharing the land with the vertebrates, and already very diverse, were the insects. The rise of the angiosperms set off one of the most dazzling coevolutionary stories ever, that of flowers and the insects that pollinate them. To start with, beetles were probably the main pollinators of the flowers whose fossils have been found in Cretaceous strata, as they still are today among certain primitive flowering plants. It is not known if pollinating bees and butterflies had evolved by the end of the period, though they were certainly in existence by early Tertiary times.

In the Seas

Meanwhile, in the oceans, the diversification of the planktonic flora was matched by the explosive evolution of new floating microscopic animals, notably certain foraminiferans. These single-celled animals with tiny chambered shells mostly inhabit the seafloor, but one group with globular calcitic shells—the globigerinaceans—evolved planktonic habits in the Jurassic; in the late Cretaceous they diffused widely through the world's oceans, leaving a comprehensive fossil record of their evolution as their dead shells fell to the ocean floors.

Many animals lived among the floating larder furnished by the plankton. Still abundant from the Jurassic were the ammonites and belemnites, though from late Jurassic times onward the differentiation between the Tethyan species and those occupying the northern, "Boreal" seas of northern Europe, Russia and Siberia and the North Atlantic became pronounced. In the latter part of the Cretaceous, the belemnites virtually disappeared from Tethys, though continuing to flourish in Boreal waters and to a lesser extent in the southern seas flanking the disintegrating Gondwanaland. In addition to climate, the partial isolation of seas also caused differentiation by promoting the evolution of "endemic" species—those that did not find their way into other regions. The "Western Interior Seaway," occupying much of central North America, for example, had many endemic ammonites toward the end of the Cretaceous.

Although the belemnites probably lived much like the

► ▼ **Some examples of late Cretaceous animal life.**
(1) *Agrypnus*, a genus of beetle. Body length 1·5cm (0·6in).
(2) *Cupes clathratus*, a species of beetle. Body length 1·5cm (0·6in).
(3) *Deltatheridium*, a genus of the early placental, insectivorous mammal family Deltatheridae. It probably fed at twilight. Length including tail 45cm (18in).
(4) *Zalambdalestes*, a genus of the family Leptictidae. It probably lived on the forest floor and had a diet of worms, beetles and small insects. Length including tail 51cm (20in).

predatory squids of today, experimental work on models of ammonite shells shows that even the most streamlined species with narrow discoidal shells would have been poor swimmers compared with most open-water fish and squids, for example. Certainly, most of those with uncoiled shells could barely have swum at all. So, although some ammonite species may have preyed on other large animals, catching them in their squid-like tentacles, most probably fed on the microscopic organisms and organic detritus that floated around them or that lay on the seafloor. However, theirs was not necessarily a quiet life: some ammonite fossils from the Upper Cretaceous of North America, for example, show the marks of having been bitten by some huge marine lizards called mosasaurs. Yet for these "sea-dragons" the ammonites were probably only an occasional snack. Along with the predatory marine reptiles surviving from the Jurassic—the dolphin-like ichthyosaurs and the swan-necked paddling plesiosaurs as well as the crocodiles—they probably largely fed on the fish of the period, particularly the abundant teleosts. Originating in the late Jurassic, this most advanced group of fish rapidly diversified throughout the Cretaceous, while the older groups of bony fishes declined. Reduction of the scales and modification of the fins allowed teleosts an astonishing repertoire of deft swimming maneuvers, which, combined with a new mobile upper-jaw system, had opened up many novel ways of feeding to them. But the dominance of teleosts was not total: their ancient relatives, the sharks and rays, also evolved several new stocks at this time.

While these evolutionary events went on overhead, so to speak, a complex of interrelated changes affected life on the seafloor, referred to collectively as the "Mesozoic Marine Revolution." Many new predators with specialized dietary preferences arose, particularly forms adapted to feed on various shelly prey. For example, from mid-Cretaceous times onward several new families of predatory gastropods evolved, variously capable of drilling through or grinding away the margins of the shells of their prey, or of asphyxiating them by smothering them. Many new crabs and lobsters, which tend to smash or snip apart the shells of their prey, as well as shell-crushing fish also arose at this time. Meanwhile the intensity of grazing on bottom-living algae was also being stepped up as yet other groups of gastropods, sea urchins and various teleost fish diversified. Some of these behaved like "biological bulldozers," dislodging any static animals attached to the seafloor that lay in their paths. There was consequently a boom in animals that avoided at least some of these dangers of life at the surface by actively burrowing well down into the sediment. Prominent among these were bivalves with extended siphons for filter-feeding from the overlying water, and the "irregular" sea urchins, feeding from the sediment itself. Some other surface-dwellers continued to thrive by virtue of increased mobility, and the ability to resettle, such as the scallops and stalkless crinoids, or by thickening or improved armoring of the shell, as in several gastropod and bivalve groups. In contrast, some of the older stocks such as the brachiopods and stalked crinoids, that characteristically lived rooted to the seafloor with little ability to resettle, and which had been so successful in the Paleozoic, made little recovery from the reduced levels of diversity reached in the Permian and Triassic mass extinctions.

Though some authorities identify a trend whereby the new forms appear to have evolved in inshore environments and then spread out over the marine shelves at the expense of the older stocks, there also seems to have been some geographical control on the pattern of changes of the marine revolution. Most of the predatory gastropods of the mid-Cretaceous, for example, were inhabitants of Boreal waters, while many species of immobile surface-dwelling animals continued to thrive in shallow Tethyan shelf seas right until the very end of the Cretaceous. Prominent among the latter were some bizarrely-shaped bivalves called "rudists." Their thickened, elongate shells grew attached to, or lying upon, the seafloor, and some species even built up sizable reef-like mounds both in quiet as well as in more turbulent water. Several other surface-dwelling thick-shelled animals, particularly other bivalves, gastropods and some "giant" (up to thumbnail-sized) foraminiferans, lived in association with these. Though corals were also present, they played relatively little part in reef formation, in contrast to their descendants of today.

Both on land and in the sea, then, the Cretaceous was a period of rising faunal diversity. In part this was due to the increasing fragmentation of the continents and seas, which

3

▲ ▶ **A late Mesozoic seafloor scene.** (1) A belemnite. (2) *Hoplopteryx*, an early spiny finned fish (family Berycidae, order Beryciformes). (3) An anemone. (4) An ammonite. (5) A crab peeling open a gastropod. (6) Algal fronds. (7) A detritus-feeding bivalve. (8) A predatory gastropod. (9) A sea urchin. (10) Bivalve shells. (11) A sea urchin moving through sediment. (12) Bivalves with siphons.

◄ **Seafloor mollusks of the late Mesozoic era.** (1) A reef-building mollusk (family Hippuritidae). (2) Family Radiolitidae. (3) A reef-building mollusk (family Requieniidae).

▼ **Closely related to herrings** is the extinct genus *Diplomystus*. It flourished in the late Cretaceous and until the middle Eocene, surviving the period of extinctions at the end of the Cretaceous. Length 50cm (20in).

promoted the proliferation of endemic species. The high sea levels of the period also played a part, in both providing vastly increased areas of shallow seafloor for exploitation by the marine fauna, and promoting equable maritime climates across the continents, which allowed the development of rich ecosystems with many species showing specialized life habits.

The close of the period was then marked by an extraordinary episode of mass extinction, which swept away some of the major components of Cretaceous life such as the dinosaurs and pterosaurs on land, and the great marine reptiles, the ammonites and the rudists in the seas, leaving open vast new ecological opportunities for the rootstocks of today's fauna to exploit in the succeeding era.

What Brought the Mesozoic World to an End?

The end of the Cretaceous was marked both by a mass extinction—an episode of significantly increased extinction rates—and by some major physical changes. Besides the crucial reorganization of seas and continents and the rapid fall in sea level of this time, there was also a drastic occurrence that left a sort of chemical signature right at the boundary between Cretaceous and Tertiary strata in those few sequences which were not subjected to erosion by the sea-level fall. Following the detection of this signature in 1978, two questions are now hotly debated: firstly, what exactly was the mysterious "Boundary Event"; and secondly, was it the sole cause of the mass extinction, or just one of many contributory factors?

The chemical signature consists of a considerable enrichment (relative to the strata above and below) in certain heavy elements, most notably iridium. These are rare in the earth's crustal rocks, though they are much more abundant in certain meteorites. The most popular explanation nowadays is that a large asteroid or comet (perhaps about 10km, 6mi, in diameter) collided with the earth, throwing up much dust and vapor

which was then pumped up into the stratosphere as the atmosphere collapsed back into the hole punched through it. This ejected matter soon shrouded the earth, causing months of nearly total darkness and perhaps several years of gloom as the dust gradually settled out, forming a universal blanket with the distinctive signature of the original impacting body. The resulting suppression of plant photosynthesis is blamed for the mass extinction. Other explanations for the signature are possible, such as a concentration of normal background meteoritic debris (which falls all the time), brought about by an episode of nondeposition or even removal of the fine oceanic sediment that dilutes it at other times. Such an effect might well have accompanied the fall in sea level. Yet the presence of the

Physical and Floral Background

Today's geography first began to take shape in the Late Mesozoic. The movements over the globe of the continents spawned by the breakup of Laurasia and Gondwanaland started the gradual jawlike closure of the equatorial Tethyan Ocean in the east, while, to the west, step-by-step separation of the Americas from Africa and Europe eventually created the Atlantic Ocean.

The opening central Atlantic was fully connected with Tethys by the late Jurassic, allowing the westward flow of water through the resulting sluice between Gondwanaland and Laurasia. Throughout the Mesozoic, rifted micro-continental fragments from northeastern Gondwanaland were transported northward by oceanic plate movement and plastered onto southern Asia. Then Gondwanaland itself began to break up in the late Jurassic and early Cretaceous, as the continents flanking southern Africa drifted away from it, creating the Indian and South Atlantic oceans. The subsequent northward drift of Africa/Arabia and India was eventually to close Tethys. As South America rotated away from Africa, the South Atlantic encroached northward between them, and in mid-Cretaceous times its surface waters broke through to the central Atlantic.

Meanwhile, northward extension of the latter continued as North America separated from Europe, creating the North Atlantic.

The opening of all these new oceans was accompanied by the rising-up of new mid-oceanic ridges, which displaced water from the ocean basins, causing the world sea level to rise. So the continents became progressively inundated by broad, shallow seas. This was enhanced in the late Cretaceous by extensive volcanism and uplift of the ocean floor in parts of the Pacific, on a scale never subsequently repeated. Estimates vary, but the peak of sea-level rise towards the end of the period may have been from 300m to over 600m (1,000–2,000ft) above the present level, drowning some 40 percent of the continental areas.

The spread of shallow seas promoted the absorption of solar heat and so climates were warmer than at present, with latitudinally broad climatic belts. Furthermore warm oceanic currents may have ranged into high altitudes. It is unlikely that there were polar ice caps like those of today (although some authorities dispute this). Nevertheless, the drowning of the continents does seem to have provoked increasingly humid conditions, with widespread monsoonal

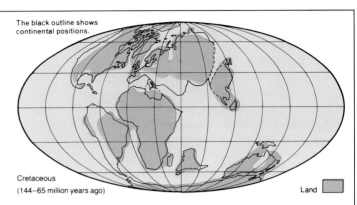

The black outline shows continental positions.

Cretaceous (144–65 million years ago)

Land

weather patterns, during the Cretaceous. In contrast, much of the late Jurassic world had experienced arid climates.

The late Mesozoic also witnessed major floral changes. On land the angiosperms (flowering plants) evolved in the early Cretaceous. Initially, the forests were dominated by conifers, cycads, gingkos (maidenhair trees) and various other woody forms, as well as the ubiquitous ferns. By the close of the period the angiosperms were already dominant, giving the forests a distinctly modern look. However, there was still no grass. In the sea plankton was greatly enriched by the evolution of many new species of single-celled algae. Some of these bore minute plates of calcite, called coccoliths, which on sinking accumulated over huge areas of seafloor as deposits of ooze. These now form the chalk of, for example,

northern Europe and Texas.

Perhaps the most extraordinary part of the late Mesozoic was its ending. Plate movements were at a critical phase as parts of the Tethyan seaway became tightly bottlenecked by collisions between Eurasia and promontories of the African/Arabian continent. Meanwhile, there was by now deep-water circulation between the North and South Atlantic, and to the north a marine connection with the Arctic basin had formed via the Labrador Sea. Rapid sea-level fall at the very end of the period exposed many previously marine areas to erosion.

Finally, there may also have been a catastrophe caused by a large extraterrestrial body colliding with the earth. Not surprisingly, the close of the Cretaceous was marked by a mass extinction, both on land and in the sea.

signature in some continental sequences, as well as that of mineral grains of possible impact origin in North America, weighs in favor of the impact hypothesis.

The timing and causation of the mass extinction are more problematical. The Boundary Event can certainly be blamed for some of the most drastic extinctions, particularly among the oceanic plankton; possibly only one species of planktonic foraminiferans, for example, survived to restock the oceans in the Tertiary. However, at one of the fullest sections, in Tunisia, many coccolith species characteristic of the Cretaceous disappear from the sequence not precisely at the boundary, but in the few metres of sediment above it, representing some tens of thousands of years of deposition. So an "instant death" explanation for the extinctions (due to poisoning, prolonged blackout or heat shock, for example) can be rejected. Rather, erratic fluctuations of climate, presumably sparked off by the event, were probably responsible.

Correlation between the marine and continental sequences is very difficult, and it has yet to be shown conclusively that the mass extinctions in the two occurred at the same time. Indeed the dinosaur extinctions may have been at different times in different places, again implying some climatic control.

Some extinctions also preceded the Boundary Event. Ammonite and belemnite genera, for example, declined rapidly over several million years prior to the end of the Cretaceous, probably because of less dramatic causes.

Finally, different groups of organisms were very unevenly affected. For example, the angiosperms and the mammals on land, and the bivalves (other than rudists) and the teleosts in the sea, continued diversifying with little setback. This would suggest a multitude of subtle causes based on changing interactions in collapsing ecosystems, more than a single indiscriminate physical cause for the extinctions.

The overall pattern now emerging, then, seems to comprise a possible extraterrestrially-provoked catastrophe coming in the midst of what was already a time of ecological crisis. But it will be some years before the dust settles on this debate and a really clear picture of what happened emerges.   PWSK

TERTIARY LIFE

The diversification of mammals: placentals in the north and in Africa; marsupials in the south. . . The evolution of hoofed mammals in conjunction with carnivore evolution. . . Primate evolution and the origin of man. . . Bird diversification. . . Reptiles and amphibians. . . Insects. . . Marine life: teleost fish, invertebrates, whales, seals, corals. . . Changes in geography, climate and vegetation

THE Cenozoic era, extending from 65 million years ago until the present day, comprises two unequal periods—the Tertiary, lasting up to about 2 million years ago, followed by the Quaternary. Successively younger strata from the two suberas contain more and more still-living species among their fossils. Early in the 19th century Sir Charles Lyell used this fact to define several successive epochs based upon the proportions of species of fossil marine shells still surviving today. The Tertiary epochs are named, from the oldest to the youngest, Paleocene, Eocene, Oligocene, Miocene and Pliocene.

With the land cleared of dinosaurs, the mammals emerged from their cryptic woody haunts and rose to dominance with spectacular rapidity. Within 15 million years (over the Paleocene and Eocene) they had given rise to a spectrum of families ranging from hoofed herbivores to clawed carnivores on land, and from bats in the air to whales in the oceans. The dispersal of the Tertiary continents lay behind this evolutionary exuberance. They formed, so to speak, several more or less separate evolutionary crucibles in which many species arose in isolation, later spreading to other areas as and when the ever-changing geography allowed.

The northern continents of Eurasia and North Africa were particularly productive in the Paleocene and Eocene, with, for example, the ungulates, various placental carnivores, bats, rodents, hares and rabbits arising alongside the primates, insectivores and the initially very successful condylarths (primitive ungulates such as the pony-sized *Phenacodus*) dating from the Cretaceous, as well as several more archaic stocks including the multituberculates (stocky little herbivores, such as the woodchuck-sized *Taeniolabis*). Until the early Eocene there was much faunal interchange between the northern continents, but this broke down in the Eocene, leading to faunal divergence, as North America finally separated entirely from Europe and as a large interior seaway temporarily developed southward from the Arctic across the Russian platform, separating the Asian from the European fauna. In the Oligocene this seaway disappeared, and a major faunal turnover ensued in western Europe as Asian forms flooded in.

The southern continental faunas followed more independent histories. Of the major groups of land mammals spreading across the globe in the Cretaceous and Tertiary, only the marsupials and the egg-laying monotremes, it appears, reached Australia prior to its separation and northward drift away from Antarctica. Consequently, the marsupials in particular diversified there through to the present day, filling many ecological niches occupied elsewhere by placental mammals. Of the latter only the bats and whales made it to the isolated Australian continent in the Tertiary, to be joined in the Pliocene by a few rodents.

The marsupials also diversified in South America, again providing both herbivores and carnivores (such as the borhyaenids). However, their rule on that island continent was shared with many widespread placental stocks, such as the edentates and notoungulates (primitive, heavy-boned ungulates, ranging up to the size of rhinoceroses, as in *Toxodon*) as well as some wholly endemic groups such as the horse-like litopterns and the bizarre astrapotheres. Many of the South American species became extinct and were replaced by North American forms following the formation of the Panamanian land bridge in the Pliocene. Nevertheless others, such as the opossums, profited from the connection and spread far up into North America.

Africa probably derived much of its mammalian fauna from the north in earliest Paleocene times, but with the isolating effect of the Tethyan seaway, several new placental stocks arose there too, including the elephants, hyracoids and embrithopods (eg the massive, horned herbivore *Arsinoitherium*) as well as the first anthropoid primates. When a land connection between Africa/Arabia and Eurasia finally became established in the early Miocene further exchange took place, with the elephants and anthropoids spreading far into Eurasia and many northern stocks such as the advanced carnivores (including cats and dogs), odd-toed ungulates such as rhinoceroses and numerous even-toed ungulates penetrating Africa.

The major climatic and floral changes of the Tertiary also contributed to the rapid evolution of mammalian faunas. In particular the pronounced change towards cooler, drier climates from the Oligocene onwards promoted the spread of grassland savannas, especially in Asia and North America. These served as nurseries for the massive proliferation of the modern grazing ungulates such as horses, rhinoceroses, camels

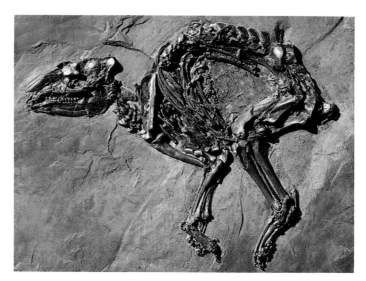

▲ **A remarkably well-preserved skeleton** of an Eocene horse, found in West Germany. The site yielded some specimens with food preserved in their guts. These horses were about the size of a terrier.

Physical and Floral Background

During the Tertiary the continents moved toward their present position as the modern oceans widened between them and the remnants of the once great Tethyan Ocean closed up.

In the north the widening Atlantic joined the Arctic basin in the early Eocene, while in the south Antarctica continued to shed its adjoining continents. Last to leave was Australia, starting in the late Paleocene; and with the collapse of the Tasman Rise in the Oligocene, Antarctica became entirely encircled by ocean. The two Americas now drifted free as continental islands until their reconnection via Panama about 3.5 million years ago. Meanwhile, crustal heating and doming over much of northeastern Africa and western Arabia caused a triple junction of rifts to form there. One gave birth to the earth's youngest ocean, the Red Sea, and another to its connecting arm to the Indian Ocean, the Gulf of Aden. The third, "failed" arm persists today as the East African Rift system.

The expansion of these ocean basins was matched by subduction of ocean floor along the margins of Tethys, which therefore closed up along a suture now preserved in the great belt of mountains stretching from the Pyrenees and the Alps in the west through to the Himalayan ranges and beyond to the Sunda Archipelago in the east. The long history of collisions involved had been heralded near the end of the Cretaceous by the pushing up of some oceanic crust over the continental margin of Arabia to form the embryonic Oman Mountains. Other collisions followed. For example, India fused to Asia in Eocene times, while a northern promontory of Africa formed by Italy and parts of the Balkan Peninsula was driven into Europe, so causing uplift of the Alps from Oligocene times onward.

Nevertheless, with only a brief interruption occasioned by the sea-level fall at the end of the Cretaceous, a corridor of shallow seas maintained communications, across continental crust, between western Tethys and the Indian Ocean right up until the early Miocene. Thereafter, the Middle East land connection between Africa/Arabia and Eurasia closed the marine corridor and allowed a north–south interchange of land animals. Two relics of Tethyan sea remained—the Mediterranean, and an expanse of shallow sea north of the suture, called Paratethys. The latter soon became filled with brackish to fresh water due to runoff from the surrounding lands, and dwindled to leave the present Black, Caspian and Aral Seas. The Mediterranean suffered an extraordinary fate: towards the end of the Miocene it became isolated from neighboring oceans and virtually dried out by evaporation (perhaps several times over, with intermittent replenishment), to leave vast deposits of salt. The period of desiccation lasted about one million years and was brought to a close by movements in the Straits of Gibraltar, which allowed refilling from the Atlantic to create the modern Mediterranean Sea.

Much ocean floor was also consumed in the Pacific, along the western coasts of the Americas, resulting in sustained uplift of the Andes and Rockies.

The equable climates of the Mesozoic degenerated in the Tertiary. The encirclement of Antarctica by cold ocean water led to glaciation, at first localized in the Eocene, but then with sea ice forming in the early Oligocene. Then, toward the close of the Miocene, the glaciation became more intense, heralding the Great Ice Ages of the Quaternary, in which the Arctic also participated.

World sea level was never again as high as it had been in the late Cretaceous, and transfer of ocean water to the Antarctic ice cap caused further erratic falls in sea level. Consequently continental seas were fewer and less widespread than in the Cretaceous, and they fluctuated in extent rather markedly with the changes in sea level and with mountain-building episodes.

With polar cooling the climatic belts progressively bunched towards the Equator, particularly from Oligocene times onwards. Thus, for example, in North America evergreen broadleaves extended to 60° north, with a warm temperate deciduous flora lying to the north in the Paleocene, and in middle Eocene times there was even paratropical rain forest in the Gulf of Alaska. But at the end of the Eocene conditions rapidly deteriorated and the deciduous broadleaves spread far southward, leaving behind them a northern conifer belt. The angiosperms nevertheless continued to diversify and dominate the land flora. Already by Eocene times they exhibited a wide range of insect and wind pollination mechanisms. A crucial event was the evolution of grasses early in the Tertiary. The extensive spread of grassland over the Oligocene and Miocene had a major impact on the evolution of mammals. The gymnosperms, however, became dominant in higher latitudes, with distinct floras developing in the Southern and Northern hemispheres.

Following its catastrophic collapse at the end of the Cretaceous, the marine planktonic flora soon picked up again with an abundance of new species of submicroscopic "nannoplankton" among such groups as the coccoliths and diatoms.   pwsk

The black outline shows continental positions.

Paleocene (65–55 million years ago)

Land

Oligocene (38–25 million years ago)

Pliocene (5–2 million years ago)

and bovids. The evolution of the horses in relation to this change has been studied in considerable detail. The early Tertiary species were terrier-sized forest-dwellers which browsed on leaves; some remarkably well-preserved fossils of these little ancestral horses have been found in a middle Eocene lake deposit in West Germany, complete with their gut contents consisting of leaves along with some grape seeds. In the Miocene they started to spread onto the savannas to graze the abundant but tough grass. As they did so their teeth evolved so as to become longer, with thickened enamel ridges which compensated for the increased wear. Meanwhile the animals also became larger and their legs relatively longer with fewer toes—all adaptations for improved galloping, presumably in response to the increased exposure to predators in their new open habitats.

The prized running ability of horses is but one product of the perennial interaction between mammalian predators and prey. It is likely that the relationship was also largely responsible for a much more general feature of mammalian evolution: measurements of brain size using casts taken from the insides of fossil skulls show a long-term trend of increase in relation to body size both in carnivores and in herbivores. Indeed those of the herbivores have tended progressively to catch up with those of the larger-brained carnivores. So, as the hunters evolved sharper wits and senses for catching their prey, it seems that the latter became ever more alert at evading them. Many other features besides intelligence were involved in this "evolutionary arms race."

The carnivores evolved a wide range of specialized offensive weapons, with some types evolving independently in several different stocks. For example, large predatory species with elongate, blade-like canine teeth, known as "sabertooths," arose several times from diverse ancestors including the marsupial borhyaenids and, among the placental carnivorous groups, the primitive creodonts as well as the more advanced felids (cat family). Swinging the lower jaw far back to give maximum clearance to the sharp upper canines, sabertooths probably felled their victim by slashing their necks and throats, causing rapid fatal bleeding. In contrast, the shorter-fanged large carnivores such as the lions of today tend to use a crushing bite on the neck, or they grip the muzzle or throat, causing suffocation. The herbivores responded with a range of defensive devices, such as the improved running ability discussed above, and, in some cases, impregnability due to sheer size, as in the elephants. A colossal Miocene relative of the rhinoceros, *Baluchitherium*, stood some 6m (20ft) high. Others evolved a variety of horns, though these were commonly also employed in competitive bouts between individuals of the same species: for herding, with hierarchical social behavior, became another common tactic for survival.

Buried in the fertile loam of these ecological interactions and geographical and climatic changes lay the roots of human evolution. The small tree-dwelling early primates had spread from their North American/European cradle into Asia and Africa by early- to middle-Eocene times. Though their bodies were primitive in many respects, their brains—and their senses—were already elaborated, as well as their grasping hands with opposable thumbs and fingers, as adaptations to arboreal life. Further diversification in Africa over the Eocene and Oligocene gave rise to the lemurs and the first anthropoids (the "higher" primates, including the monkeys, apes and man): the earliest apes yet found come from the Oligocene of Egypt.

The reconnection of Africa with Eurasia in the early Miocene opened the latter to anthropoid invasion. At the same time, however, their woodland habitats were contracting and yielding to scrub and grassland as the climate became drier. Some of the anthropoids started to evolve adaptations, such as thickly enameled elongate teeth, for foraging on the tough, fibrous vegetation that surrounded their woody homes. Middle to late Miocene fossil specimens, especially of the genus *Ramapithecus*, are known from localities ranging from the Balkans to southern Asia as well as East Africa. Unfortunately their fossil record becomes extremely patchy between 8 million years ago, in the latest Miocene, and about 4 million years ago, in the Pliocene. But it seems likely that it was during this period that some ramapithecine gave rise to the earliest fully bipedal humans, as well, perhaps, as to our close cousins the great apes (chimpanzees, gorillas and orang-utans)—a fascinating story that will be picked up later in this book.

The other great vertebrate success story on land in the Tertiary was that of the birds. Like the mammals they diversified rapidly, and indeed nearly half the orders alive today had evolved by the close of the Eocene, including such adaptive novelties as penguins, vultures, owls and a variety of large flightless birds. Some of the latter, such as the Eocene giant *Diatryma*, from North America, locally enjoyed a brief reign as top carnivores alongside the early mammals. Even after the latter had come to dominate the top carnivore niches on the ground, flightlessness, usually coupled with size increase, arose many times over among birds in isolated areas such as islands. These were generally herbivorous forms. In many cases their evolution seems to have been achieved by a considerable slowing down of their body development, though with size increase continuing as before. So the adults that resulted were really like enormous overgrown chicks. One of the most spectacular examples was a colossal moa in New Zealand, *Dinornis*, standing some 3m (10ft) high. New Zealand also seems to have been the cradle of the penguins, which spread over the higher latitudes of the Southern Hemisphere.

The fossil record of bird evolution is predictably biased in favor of the big-boned giants and the inhabitants of shores and waterways, both having a reasonable potential for preservation. Some remarkable specimens give important clues to the

▶ **Some forest-dwelling mammals** of the Tertiary period (65–2 million years ago). (1) *Barylambda*, an early ungulate which probably browsed on leaves. Late Paleocene. Length about 2·4m (8ft). (2) *Ptilodus*, a rodent-like multituberculate and probably one of the first herbivorous mammals. Late Paleocene. Length about 36cm (14in). (3) *Plesiadapis*, an early lemur-like genus but retaining claws rather than nails. Paleocene. Length including tail 51cm (20in). (4) *Protictis*, an early carnivore which probably preyed on small mammals. Paleocene. Length about 90cm (36in).
(5) *Meniscotherium*, an early ungulate (condylarth). Late Paleocene. Length including tail about 61cm (24in). (6) *Palaeoryctes*, a small insectivorous mammal. Paleocene. Length including tail about 18cm (7in). (7) *Prodiacodon*, an early insectivore, probably similar to the modern hedgehog in habit and diet. Paleocene. Length about 18cm (7in).

evolutionary relationships. For example, *Presbyornis*, from early Eocene sediments of western North America, has the head and bill of a duck on a flamingolike body and is a possible common ancestor of both of these.

The proliferation of so much scurrying warm flesh offered clear rewards to any furtive "cold-blooded" predator that was particularly sensitive to their warmth and movements. Such exactly were the snakes, still diversifying to this day in tandem with their favorite dinners such as the rodents and other small mammals, frogs and birds. These, in conjunction with the still widespread lizards, crocodiles and turtles—all represented by Tertiary fossils—show that reptilian history is anything but a closed book. Early Tertiary fossils testify that the modern frogs, toads and newts were already well established by then.

Though insect—and particularly beetle—species overwhelmingly outnumber those of all other land animals today, their fossil record is patchy because of the special conditions required for their preservation. Even so, such fossil insect assemblages as there are from the Tertiary confirm the enormous diversity of the group throughout the period. For example, the oil shale lake deposits from the middle Eocene of West Germany with the exceptionally preserved horses, discussed earlier, also contain abundant insects, including beetle cuticles with their original coloration, and the wings of primitive moths in the stomachs of bats. Similarly, laminated organic-rich lake deposits of middle Miocene age in Nevada, USA, contain a wide range of insects, including beetles, wasps, bees, moths, butterflies, gnats, bugs and many others, in addition to leaves, fruits, pollen, fish, mollusks and algae. Another source of beautifully preserved insects is amber accumulated

▶ **Some large mammals of the Oligocene** epoch in North America.
(**1**) *Poëbrotherium*, an early genus of camel, about the size of a sheep.
(**2**) *Merycoidodon*, an early oreodont ruminant, about the size of a sheep.
(**3**) *Brontotherium*, a genus belonging to the superfamily of titanotheres (a group closely related to the horses); a plant-eater. Maximum height at shoulder about 2.5m (8ft). (**4**) *Daeodon*, a genus of giant pig or entelodont, an early even-toed ungulate or artiodactyl. Maximum skull length about 1m (3.3ft). (**5**) *Metamynodon*, an early rhinoceros or amynodont. Maximum length about 4.5m (15ft). (**6**) *Hyaenodon*, a genus belonging to the hyaenodonts, the main group of flesh-eating placental mammals. (**7**) *Archaeotherium*, a genus of giant pig or entelodont; about the size of a modern cow. (**8**) *Mesohippus*, an early horse, adapted for running in open country. Height about 65cm (25in). (**9**) *Leptomeryx*, an early pecoran, antedating the development of horns; a leaf-browser. Length about 60cm (2ft). (**10**) *Amphicyon*, a genus belonging to the family Amphycionidae. Its members had canidlike features but were ancestral to bears.

from pine trees, particularly famous examples coming from Oligocene deposits along the Baltic coast.

As on the land, the fauna of Tertiary lakes and rivers soon acquired a "modern" character as the freshwater mollusks, worms, arthropods and other survivors from the Mesozoic were joined by numerous new teleosts such as carps, pike, perch and cichlids, in place of the archaic (largely nonteleost) forms that had been prevalent in the Mesozoic. They were but one component of the still expanding teleost empire in the seas and oceans. By middle Eocene times all major teleost groups had

evolved, ranging from streamlined predators and plankton filterers in open water to grazers, detritus-feeders, predators and scavengers at the seafloor. These spread into all latitudes and water levels, including the oceanic depths, as well as into freshwater habitats.

The oceanic plankton rapidly rebounded from its crash at the close of the Cretaceous. New stocks of globigerinacean

◄▼ **A coral reef scene** of the Tertiary period. (1) Queen angelfish (*Holacanthus ciliaris*). (2) Longnose butterfly fish (*Chelmon rostratus*). (3) Rock cod (*Epinephelus adscensionis*). (4) Imperial angelfish (*Pomacanthus imperator*). (5) Fan coral. (6) Coral (genus *Tarbellastraea*). (7) Coral. (8) Green algae (class Dasycladacaea). (9) Coral. (10) Domed corals. (11) Coral. (12) Coral (genus *Porites*). (13) Red algae.

(14) Scallops, showing tentacle fringes. (15) A lobster feeding on a gastropod. (16) Scallops. (17) Mussels. (18) Fanworms. (19) Limpets. (20) A sea snail. (21) Dead coral. (22) Coral (genus *Tarbellastraea*). (23) Hermit crabs. (24) Crabs opening gastropods. (25) Nummulites (foraminiferans). (26) Alveolins (foraminiferans). (27) A sea worm. (28) A sea snail feeding on clams. (29) A whelk feeding on clams. (30) Red algae. (31) A bivalve mollusk. (32) Cockles. (33) A tusk shell. (34) A bivalve. (35) A lugworm.

foraminifera flourished from the Paleocene onwards, alongside myriad tiny swimming crustaceans and, from Eocene times onward, even some specialized planktonic gastropods, as well as the ever-present hosts of microscopic larvae of many different marine organisms. Among the larger swimmers the teleosts shared this domain with squids and cuttlefishes, now the dominant cephalopods following the demise of the ammonites and belemnites. The rich harvests of the open seas also lured many mammalian stocks into marine modes of life. Primitive but already huge predatory whales were in existence as early as the middle Eocene. Most of these became extinct by the early Miocene, to be replaced by the modern whales, both predatory and filter-feeding. Other new arrivals in the early Miocene of the Pacific were the seals.

On the seafloor the new scheme of ecological relationships established by the marine revolution of the late Mesozoic was rapidly fleshed out by increasing numbers of species, especially in the tropics, so that the intensity of predation on shelly prey, for example, had already reached modern levels by Eocene times. In higher latitudes the increasing seasonality ushered in by the cooler climates of the later Tertiary favored species with rather broad feeding habits that could cope well with fluctuating food sources, and it was here that the opportunistic whelks rose to prominence in late Miocene times. Other important new additions in the Tertiary included the sand dollars, a major group of burrowing irregular echinoids which first appeared in the Paleocene, and, in the early Eocene, the balanid barnacles, which dominate many rocky coasts today alongside and in competition with their slightly older cousins, the chthalamid barnacles.

As on land, the faunas of the Tertiary seas and oceans underwent much geographical partitioning. When the African/Arabian–Eurasian land bridge formed in the early Miocene, for example, the benthic faunas of the Mediterranean relic of Tethys and the Indian Ocean markedly diverged. A similar divergence of the Caribbean and Pacific faunas of Central American waters followed the emergence of the Panamanian land bridge between the Americas in the Pliocene.

One ecosystem much affected by these provincial changes was that of the tropical reefs. These were widely scattered in the early Tertiary, with many coral genera ranging from the Indo-Pacific region via the Mediterranean Tethys across to the Caribbean. But with the contraction of the Mediterranean and hence of its reefs in the Miocene, the Indo-Pacific and Caribbean reef provinces became separated. With the arrival of northern Australia in tropical waters by this time, the Indo-Pacific became the dominant region of reef development. Virtually all the new genera of corals which evolved from the late Miocene onward did so in the mosaic of barrier, fringing and atoll reefs of that region. As a result the reefs of the Caribbean today have far fewer coral genera than those of the Indo-Pacific.

As the great equatorial current system that had flowed through Tethys until the Miocene and between the Americas until the Pliocene thus became increasingly cut off, and the climates of higher latitudes began to cool, so the great gyral currents that stir the waters of the Pacific and the Atlantic intensified. The world stood on the brink of the Great Ice Age of the Quaternary period.   PWSK

QUATERNARY LIFE

Climatic changes: the geological evidence of periods of glaciation interspersed with warm periods. . . Consequent changes in the distribution of animal species. . . Mass extinction in the last 20,000 years: the possible causes. . . The last half million years: a period of adaptation to dramatic climatic changes

THE Quaternary period covers approximately the last 2 million years of geological time, up to and including the present day. The term "Pleistocene" is sometimes used with the same meaning as "Quaternary," but is more commonly used to mean the earlier epoch of the Quaternary, excluding the last 10,000 years. The Holocene is the later epoch, from 10,000 years ago to the present.

Quaternary Climatic Changes

The Quaternary is characterized by marked climatic oscillations, on a time scale of thousands of years, superimposed on a long-term trend of global cooling. These climatic changes were reflected in the geological record in a variety of ways, including the nature of the sediments deposited, profound shifts in the geographical distributions of plants and animals, and indirectly by the proportions of oxygen isotopes incorporated in the shells of microscopic marine animals (foraminiferans).

The sediments on the deep ocean floor preserve long continuous records of Quaternary events, whereas terrestrial sequences of sediments are usually fragmentary and more difficult to interpret. Fossil foraminiferans (microscopic shelled Protozoa) in the cores taken from these deep-sea sediments record Quaternary climatic changes in two distinct ways. First, changes in the proportions of different species characteristic of warm or cold waters can be used to construct graphs of changing ocean temperatures (at different depths according to species) for many localities throughout the world's oceans. Where reliable absolute dating methods are available, these data can be used to construct world maps of ocean temperatures at various times in the past.

Secondly, the ratio of the oxygen isotopes ^{18}O to ^{16}O extracted from sea water by foraminiferans in building their calcium carbonate shells (tests) varies both according to the water temperature and to how much of the world's water is locked up in glaciers and ice sheets at any one time—the "global ice volume."

The curves for oxygen isotope fluctuations, which match remarkably well for different parts of the world, record at least seven high ^{18}O peaks within the last 700,000 years. These peaks are comparable with the one at about 18,000 years ago, which corresponds to the last major glaciation recorded in the terrestrial record. Only four such glaciations have so far been recognized on land, perhaps because the deposits left behind by earlier glaciations have been largely obliterated by their successors.

The widespread occurrence in the Northern Hemisphere of a distinctive type of sediment called "till," or more descriptively "boulder clay," records the former presence of ice sheets during at least four intensely cold phases within the past 700,000 years. Similarly fossil ice wedges, ice-filled cracks in the ground

found in modern arctic regions under conditions of permanently-frozen subsoil (permafrost), are common in now-temperate regions of Europe and North America. The fossil remains of such animals as reindeer (caribou) and lemmings show that in cold phases species now limited to the Arctic ranged as far south as southern France and Texas, while conversely the remains of temperate animals such as hippopotamus in northern England date from interglacial phases, with climates as warm as today or a little warmer.

Much important information on past climates and vegetational conditions is provided by fossil pollen. The flowers of

I

▼ **Some large mammals** that flourished in northwest Europe in the last cold stage of the Quaternary period (about 40,000–30,000 years ago). (**1**) Woolly mammoths (*Mammuthus primigenius*), the dominant mammoth of the Quaternary period. It survived severe winters by building up fat. (**2**) Woolly rhinoceroses (*Coelodonta antiquitatis*), a common rhinoceros of the period. (**3**) Grizzly bears (*Ursus arctos*), then common throughout Europe where they had superseded the Cave bear (*Ursus spelaeus*). (**4**) The horse *Equus ferus*, one of several species found in Europe during the last ice age. (**5**) A lion (*Panthera leo*). Lions were widespread in Europe.

higher plants produce pollen which is dispersed by the wind, and is incorporated and preserved in the accumulating sediments of lakes, rivers and peat bogs. Careful analysis of the microscopic pollen grains reveals the nature of the vegetation cover at various times in the Quaternary; for example, in northwest Europe the alternation of open herb vegetation and temperate forests records the alternation of cold and warm climatic conditions.

Although both the marine and terrestrial sequences record the same general pattern of Quaternary climatic changes—long cold periods interrupted by relatively short warm phases of varying duration and intensity—the two types of sequence cannot be easily related to one another at present.

Faunal Changes

The complex climatic changes during the Quaternary resulted in spectacular shifts in the geographical distributions of land plants and animals. The relationships between climatic oscillations and faunal changes are best seen in western Europe, not only because this region has been better studied than elsewhere, but also because its proximity to the North Atlantic maximizes the contrasts between cold and temperate phases and their faunas.

Throughout the first two-thirds of the Quaternary, faunal change proceeded at a moderate pace, but from about 700,000 years ago, as the climatic oscillations became stronger, the effects on the fauna increased considerably. For example, during the last cold period, from approximately 40,000 to 14,000 years ago, a predominant "steppe-tundra" vegetation in western Europe was accompanied by a rich fauna, comprising: species now restricted to the Arctic, eg reindeer, lemming, Arctic fox, Musk ox; species now found in the Eurasian steppe, eg horse, Saiga antelope, sousliks (ground squirrels), jerboas, pikas; species still present in the region, eg Gray wolf, Brown bear; and extinct species, eg mammoth, Woolly rhino, an extinct bison (*Bison priscus*) and Giant deer (*Megaloceros giganteus*). Lion and Spotted hyena, also present in these faunas, were not greatly affected by climatic conditions, provided there was abundant game on which to prey.

Broadly similar faunas at this time extended across Eurasia, from the Pyrenees to eastern Siberia, and with modification across "Beringia" into North America. These assemblages are of particular interest as, even allowing for the extinct forms, they represent communities with a mixture of species which cannot be matched anywhere at the present day. Similar

◄ **Some large mammals** that flourished in southwest North America (Rancho La Brea) in the last cold stage of the Quaternary period (about 30,000–20,000 years ago). (1) *Equus*, the modern genus of horse, descended from the first one-toed form, *Pliohippus*. Horses later became extinct in America; *Equus* survived in the Old World. (2) Imperial mammoths (*Mammuthus imperator*), one of the largest proboscids (order Proboscidea) that lived on the Great Plains of southern North America. Height about 3.5m (12ft). (3) A saber-tooth cat (genus *Smilodon*). It probably preyed upon elephants and mastodons. (4) Giant ground sloths (genus *Megatherium*), the largest of all ground sloths. Maximum length 6m (20ft). (5) *Capromeryx*, a genus of four-horned prongbuck, common in North America throughout the Tertiary. (6) *Bison antiquus*, a genus that was larger than modern bisons. (7) Dire wolves (*Canis dirus*), a species similar to modern wolves. (8) *Homotherium*, a genus of Scimitar cat. It had front limbs longer than its hind limbs, giving it a semierect stance. It was probably adapted for preying on young elephants and mastodons.

assemblages of animals are known from previous cold phases.

Terrestrial mollusk faunas from these cold phases are characterized by an absence of southern species and the dominance of open grassland taxa (*Vallonia, Pupilla*). Work on beetle faunas from Britain within the range 45,000 to 10,000 years ago has shown that there were marked and sometimes rapid climatic changes. Arctic faunas are succeeded by temperate faunas at about 43,000 years ago, surprisingly suggesting summer temperatures rather higher than today, although this was a predominantly cold period.

During interglacials, in contrast, temperate animals accompanied the spread of broadleaved forests to western Europe. For example, the faunas of this area in the middle of the last interglacial, about 120,000 years ago, included Fallow deer, hippopotamus, the extinct Straight-tusked elephant (*Palaeoloxodon antiquus*) and an extinct rhino (*Dicerorhinus hemitoechus*).

The faunas of previous interglacials are of broadly the same character, but differ in details of the species present, which allows them to be distinguished from one another. Further eastward into Asia the differences between cold and temperate faunas become more blurred, as steppe-like faunas appear to have persisted throughout.

At times during interglacials summer temperatures were a little higher than now, allowing certain southern species of plants and animals to spread much farther north. For example, the European pond turtle (*Emys orbicularis*), which needs warm, sunny summers for its eggs to hatch, is known from finds in England and southern Scandinavia more than 300km (200mi) beyond its present-day range. Similarly a Mediterranean species of dung beetle (*Onthophagus opacicollis*) and southern species of mollusk (eg *Corbicula fluminalis*) are known from the middle of the last interglacial in England.

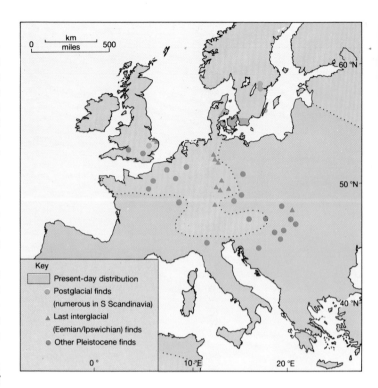

Key
- Present-day distribution
- Postglacial finds (numerous in S Scandinavia)
- Last interglacial (Eemian/Ipswichian) finds
- Other Pleistocene finds

Late Quaternary Mass Extinctions

The modern world is zoologically impoverished, because many large and fascinating animals have become extinct in the very recent geological past. The late Quaternary faunas of Europe and northern Asia about 20,000 years ago included Woolly mammoth, Woolly rhino, a bison and Giant deer—all of which are now entirely extinct—together with animals such as lion, Spotted hyena and Musk ox which still survive outside the area. Similarly the North American late Quaternary faunas contain many extinct species, including a saber-tooth cat *Smilodon*, the giant armadillo-like *Glyptodon*, ground sloths such as *Megatherium*, extinct species of camels and peccaries, the mastodon *Mammut americanum* and, surprisingly, the horse— an animal which survived in the Old World, but became extinct in the New. (The so-called "wild horses" or mustangs of the USA are feral; that is, descended from escaped domestic stock.)

Australian extinct mammals include giant kangaroos, the giant wombat-like *Diprotodon* and the marsupial "lion" *Thylacoleo*.

Available radiocarbon dates indicate that these animals perished in a mass extinction within the period of about 20,000 to 9,000 years ago. These mass extinctions represent an enormously increased extinction rate compared with those found during the earlier parts of the Quaternary, and unlike the previous extinctions were not accompanied by ecological replacement of the animals that were lost. A feature of these mass extinctions which makes them unique in the geological record is that they affected large terrestrial mammals almost exclusively. Although it was a global phenomenon, different parts of the world were affected very differently; the extinctions were especially marked in North and South America, but less so in Europe, while very few species appear to have been lost in Africa during this period.

Two major hypotheses have been advanced to account for

Climatic Changes and the Environment

The last 500,000 years or so of the Quaternary witnessed dramatic changes in climate which in turn profoundly affected both the physical and biological environments.

On at least three distinct occasions ice sheets up to 3km (2mi) thick—comparable with the present-day Greenland and Antarctic ice caps—covered large areas of North America (most of Canada and the northern USA) and northwestern Eurasia (Scandinavia, the British Isles and western Siberia). These major phases of glaciation, however, lasted for only a few thousand years each. For most of the last 700,000 years these areas were largely ice-free, with climates markedly cooler and drier than now. For much of this time these were regions of permanently frozen ground (permafrost) as in present-day arctic regions.

For most of this period the vegetation of now-temperate regions, such as northwest Europe and the eastern USA, was open "steppe-tundra," comprising grasses, sedges and other herbs with few or no trees. Warm, interglacial phases, each lasting 10–15,000 years, were accompanied by the development of temperate broadleaf woodland in these areas. The world is now living within the later part of such an interglacial phase, and excluding the uncertain effects on future climates of man-made atmospheric pollution—especially the large increase in carbon dioxide from fossil fuels—colder conditions should return within the next few thousand years.

Recent research has shown that tropical regions of the world also saw important changes in climatic and vegetational conditions during the later part of the Quaternary. The "steppe-tundra" phases of northern Eurasia and North America were paralleled by phases of increased aridity in, for example, tropical South America, where the rain forests were drastically reduced and fragmented at the expense of grasslands. Conversely during warm phases the forests expanded and replaced the grasslands.

Previously it was widely supposed that the tropical regions had remained essentially stable for hundreds of thousands, even millions of years. It is now clear, however, that no part of the world has escaped the effects of the Quaternary climatic changes, and that everywhere plants and animals have had continuously to adapt to these changes, or else become extinct, as many did.

The changes in climate were accompanied by large changes in sea level. At times of major glaciation the vast quantities of global water locked up in ice sheets resulted in world-wide falls in sea level of 100m (330ft) or more. During interglacials, following the melting of much of this ice, sea levels rose and were as high as today's, or sometimes higher.

When sea levels were low, many areas now flooded by shallow seas were connected by land bridges. For example, the Bering land bridge ("Beringia") connected Asia to North America via eastern Siberia and Alaska; New Guinea was connected to Australia; and Britain was joined to continental Europe across the dry bed of what is now the southern North Sea.

These land bridges were of great importance in the Quaternary as they allowed the migration of plants and animals between areas now separated by stretches of salt water, now impassable to most terrestrial species.

---

this phenomenon: that the extinctions were caused by climatic changes; or that they were due to excessive hunting—"overkill" by prehistoric man. The proponents of the first hypothesis point to the marked and rapid changes in climate and vegetation that occurred toward the end of the last cold stage. Over large areas of northern Eurasia and North America the open "steppe-tundra" biome with its rich fauna was quickly replaced by forests. Some species, including reindeer, Arctic fox and lemmings, found refuge in present-day arctic regions, while others, including Saiga antelope, horse and sousliks in Eurasia, survived in the steppe grassland. A third group is supposed to have been unable to adapt to the changing environment, and so became extinct. In many parts of the world, such as Australia and the American southwest, extinctions have been attributed to a marked postglacial increase in aridity.

A major argument against the climatic hypothesis is that similar environmental changes appear to have taken place at the close of previous cold periods, without resulting in mass extinctions. Supporters of this hypothesis, however, suggest that such animals as mammoth only acquired specialized adaptations to "steppe-tundra" during the last cold stage.

The alternative "overkill" hypothesis, however, is also unsatisfactory in a number of important respects. It was originally suggested that the late arrival of man in the New World about 12,000 years ago had a drastic impact on the mammal fauna. "Big-game" species with no previous experience of man were thought to have fallen an easy prey to the stone and bone-tipped weapons of these Upper Paleolithic hunters. There is now, however, abundant evidence for man in the New World well before 12,000 years ago; and, moreover, similar if less marked extinctions occurred in Eurasia at about the same time where man had been present from much earlier in the Quaternary. Furthermore, there is little evidence from archaeological sites that the animals that became extinct were those actually hunted, and if small populations of hunters with relatively primitive weapons were responsible for exterminating so many large mammals, it is very surprising that many others have survived, albeit precariously, to the present day in the face of far larger human populations, better hunting methods and the use of firearms.

The solution to this intriguing problem awaits the results of careful study on local, regional and global scales.   AJS

The Background to Evolution

IN the mid-19th century the subject of evolution took a giant step forward with the publication of Charles Darwin's theory of evolution through natural selection—a theory that has largely survived, with modification and development, to this day. On an expeditionary voyage around the world in 1831–36 Darwin collected together an enormous number of facts suggesting evolutionary change had taken place, and afterwards connected them by his theory of the mechanism by which the changes could have occurred. He himself was conscious of other evolutionary thinkers as far back as the time of Aristotle and Cicero, but all these predecessors, including his grandfather Erasmus, could only speculate about the causes of evolution. Darwin's theory was a truly scientific one, open to the experimental test. The story of the arguments about and eventual acceptance of Darwin's ideas is a fascinating one, continually interwoven with theological and political views and principles, and rivalry between natural historians.

◄ **Evolution in miniature.** A Marine iguana (*Amblyrhynchus cristatus*) on the Isla Santa Cruz in the Galapagos Islands. The Galapagos contain many animals related to ones found elsewhere (especially in South and Central America) but which are clearly members of distinct species. This observation stimulated Darwin's thought about the origin of species. The Marine iguana has several interesting features; because it lacks natural enemies it is unafraid of man.

DARWINIAN EVOLUTION

Why was the origin of species a problem?... Order and design in nature as observed by Cicero and, later, Henry More (1655)... John Ray (1691) finds variety as well as order in nature... Irregularities in nature suggest some lack of design... Evolution accounts for irregularities... Darwin and the theory of natural selection... Pre-Darwinian attempts to explain both the unity and the diversity in nature... Darwin develops his theory of the origin of species... Conservative objections to Darwin's theories... Support from contemporary scientists... Controversy continues... The life of Charles Darwin... A. R. Wallace: Darwin's codiscoverer... Darwin's "sexist" view of women is refuted by Wallace... Animal classification

CHARLES Darwin was far from being the first evolutionist. His achievement was that of being the first scholar (jointly with A. R. Wallace) to put forward a theory of evolution based on empirical evidence, and giving an account of the mechanism by which new species may have come into being; but evolutionary ideas had been around for quite a while. Admittedly, many of these ideas were quite vague and unsatisfactory in modern terms. Lamarck's theory, for example, that the simplest animals were spontaneously generated from nonliving matter, and that these then evolved into ever more complex forms in response to their environment, was clearly derived from a philosophical belief in steady progress.

But the fact remains that the existence of such beliefs, as an alternative to the "orthodox" view that each individual species was separately created by God to fit its intended environment, helped to prepare the ground. Darwin's ideas were met by strong opposition from conservatives, particularly certain churchmen; but they were welcomed by many who saw for the first time the concept of evolution given definite scientific form and content.

To take one step back, why was the origin of species a problem? That is to say, what led scientists before Darwin, as well as Darwin himself, to question the "creationist" account? It was mainly their growing realization that the traditional interpretation of God creating the whole universe with a six-day period was untenable. This came about as geologists began to date the earth and paleontological events on an even longer time scale and as the world's fauna and flora became better known it became apparent that many species had really become extinct (and were not merely waiting to be discovered in unexplored lands). This was not a rejection of God as such, but merely a rethinking of God's methods of working; many scientists then as now believed in God as the object of faith. The origin and existence of species had to be explained by natural laws, like all other phenomena found in the natural world.

Secondly, the idea of special creation failed to account satisfactorily for all that was known about living beings. There is indeed regularity and design in nature—a classic argument for the existence of God from the time of Cicero. Yet this regularity is not perfect. Human beings, like other organisms, possess organs (such as the appendix) that are apparently useless; such vestiges had been known since Aristotle. They suggested that species had not always been as they are now—a fact amply confirmed as soon as fossils were recognized for what they were.

On the other hand, there is *more* regularity among living creatures than can easily be accounted for by special creation. As soon as modern forms of classification were developed, with species being grouped according to their common characteristics, it became clear that different species must be related more or less closely; and it is a small step from this to see the relationship as one of descent, in which characteristics are inherited and slowly changed. Before Darwin, the mechanism of the change was obscure, but it could be seen that some such physical relationship between species was necessary to explain the many close similarities.

Darwin's great innovative step was to introduce the idea of *natural selection* as the mechanism for evolution—that the pressures of competition for a limited food supply within a hostile environment enabled only the fittest to survive and reproduce. Thus some of the many variations carried by individuals in any species will make them better adapted to breed and survive in that environment; these variations will increase in frequency, and come to predominate.

Does natural selection also apply to social and moral characteristics within human society? Concern that it might do, and a religious determination that it should not, fueled much of the opposition to Darwin—along with the uncomfortable feeling that to have an ape as an ancestor (as Darwin was caricatured as saying) was scarcely fitting for a Victorian gentleman.

Order and Design in Nature

The distinguished geneticist J. B. S. Haldane was once asked what can be inferred legitimately about the Creator from a study of the works of creation. He replied: "An inordinate fondness for beetles." (Three-quarters of the world's known animals are insects, three-quarters of these beetles—and three-quarters of these are weevils.) His answer was not merely frivolous. A study of the creation and especially of living things has been used to support atheism, agnosticism, theism, the existence of

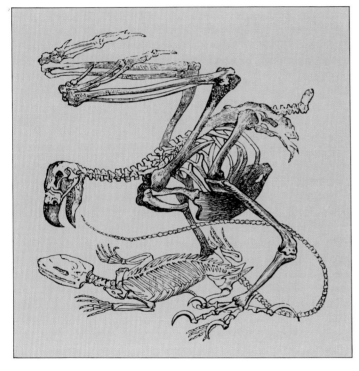

both God and the Devil, and an omniscient Creator working through a second principle which would explain the imperfections and mistakes. Probably far more people have been impressed by the design and ingenuity shown in Nature than have not, even though they have needed subsidiary principles to account for vestiges, tumors, pestilences, pain and similar phenomena.

The argument from design to the attributes of God is very ancient. Cicero summed it up for the ancient world in *On the Nature of Gods*, in passages quoted as authoritative statements of the argument into the 19th century. The argument of Archdeacon Paley's *Natural Theology*, which Charles Darwin studied and (rightly) admired for its superb exposition, is virtually that of Cicero. There is far too much regularity, contrivance, design in the world for it not to be the product of intelligence; indeed, of divine intelligence, since the works of Nature so surpass those of mankind—compare a statue of a man with a real man. Yet

the works of man are self-evidently and by our own direct knowledge the product of intelligence. Cicero uses the example of an orrery, an instrument designed to reproduce the motions of the principal heavenly bodies—if it were taken into Britain or Scythia, would not even those barbarous natives recognize it as a work of reason? How much more, then, the world, which contains both the orrery and its maker (man), and far more ingenious natural things besides!

The beauty, regularity, and uniformity of motion of celestial bodies, the marvelous construction of the human body (eg the organs of speech for speaking), and the bounty of Nature producing things for all but especially for man, are all set forth by Cicero, as well as the awefulness of storms, earthquakes and monstrosities, our innate ideas of the gods, and the validity of augury and prophecy. Objections are also raised, but not from any detailed knowledge of comparative anatomy or of the order of Nature; one objector points to hot and cold deserts as not the best of land-usage, and another urges forcibly the prosperity of the wicked, their atrocities, and pain, against the proposition that Nature could not be better contrived.

The argument from the wonderful contrivances of the natural world was borrowed by some of the Latin Fathers, and it may have influenced St Thomas Aquinas, but not until the 17th and 18th centuries did it become really influential. The Cambridge Platonist Henry More's *Antidote against Atheism* (1655) owed much to it, and like Cicero he had to argue against the idea (of Leucippus, Democritus, Epicurus and Lucretius) that the world was the result of the fortuitous congregation of atoms—the idea that matter and motion are the world's ingredients had been revived. More, speaking of the bountiful provision of plants for animals to feed on, therefore says: "But this seeming rather *necessary* than of *choice*, I will not insist upon it. For I grant that Counsel most properly is there imply'd, where we discern a variety and possibility of being otherwise, and yet the best is made choice of. Therefore I will onely intimate thus much, that though it were necessary that some such thing as *grass* should be, if there were such and such creatures in the world, yet it was not at all necessary that grass and herbs should have that colour which they have; for they might have been red or white, or some such colour which would have been very offensive and hurtful to our sight." (A conclusion which cannot now be sustained.)

John Ray, both a divine and an excellent scientist, remarks in the preface to his highly influential *Wisdom of God manifested in the Works of the Creation* (1691) that "you may hear illiterate Persons of the lowest rank of the Commonalty affirming, That they need no Proof of the Being of a God, for that every Pile of Grass, or Ear of Corn, sufficiently proves that: For, say they, all the Men of the World cannot make such a thing as one of these; and if they cannot do it, who can, or did make it but God? To tell them, that it made it self, or sprung up by chance, would be as ridiculous as to tell the greatest Philosopher so." Ray finds a strong argument in the achievement of the same end by different means. "So the infinitely wise Creator hath shown in many Instances, that he is not confined to one only Instrument for the working one Effect, but can perform the same thing by divers means. So, tho' Feathers seem necessary for flying, yet hath he enabled several Creatures to fly without them, as two sort of Fishes, one sort of Lizard, and the Batt, not to mention the numerous Tribes of flying Insects."

Ray was too good a scientist not to realize that it was difficult to attribute design to some structures, and too honest not to say so. "The Body of Man may thence be proved to be the effect of Wisdom, because there is nothing in it deficient, nothing superfluous, nothing but hath its End and Use . . . Only it may be doubted to what use the *Paps* in Men should serve. I answer, partly for Ornament, partly for a kind of conformity between the Sexes, and partly to defend and cherish the Heart; in some they contain Milk, as in a *Danish* Family we read of in *Bartholine's* Anatomical Observations. However it follows not that they or any other parts of the Body are useless, because we are ignorant." This last remark was eminently reasonable at the time.

With the development of comparative anatomy, the existence of vestiges (already known to Aristotle) became more impressive. Erasmus Darwin (1731–1802) interpreted them rightly as evidence of change of function, explicable only by evolution. But as long as puzzling cases could be put down to ignorance, or viewed, as were diseases, as God's punishments (surely sometimes excessive?), here was a complete system of explanation that did justice to the overwhelming evidence of adaptation in the living world and that could be used triumphantly against materialists, atheists and other subversives. When Newton reduced the heavens to matter and motion it was bad enough; but the divine drama needs a reliable stage to be acted on, and his ideas did not affect man. One can appreciate the feeling of utter consternation that must have overwhelmed so many intelligent and devout people when the theory of evolution (already pushed by French atheists) became a serious scientific explanation of far greater power than the argument from design.

At present the whole subject has taken on a completely different aspect; a question hotly debated—and heavily influenced by politico-religious attitudes—is how many of the evolutionary phenomena are regularities caused by natural selection, and how many purely fortuitous.   AJCa

Darwinian Evolution

Many theories of evolution have been devised to explain how living things acquired their present forms. "Darwinian evolution" refers to a particular set of theories, with their related assumptions, by which the English naturalist Charles Darwin (1809–1882) sought to explain the phenomena of life in the middle decades of the 19th century. Darwin was remarkably successful, although some of his theories and assumptions are now discredited.

Darwin was a man of "enlarged curiosity." He theorized about almost everything. As a Victorian gentleman he held certain assumptions about the natural world that gave his theories a distinctive character. For example, Darwin believed that beneficial change in nature was gradual and, on the whole, progressive. He believed that the benefits of change could be assessed by its usefulness to individuals. He believed that individual usefulness could be expressed as a ratio of pleasure to pain. ("All sentient beings have been formed so as to enjoy, as a general rule, happiness.") He believed that all

change occurred through the interaction of particles of matter according to fixed and uniform laws—a mechanical, Newtonian view of the universe. He believed that large and complex changes could be reduced to a sequence or pattern of many simple changes. All these assumptions entered into Darwin's theory of natural selection. Evolution today deserves to be called "Darwinian" to the extent that it rests on similar assumptions.

The Theory of Natural Selection

The theory of natural selection was both simple and profound. It began with two observations. First, more organisms are born that can survive to reproduce themselves, because the environment has limited means of subsistence. This is superfecundity (excessive fruitfulness). Darwin held that superfecundity results in a struggle for existence. Plant and animal species compete within and among themselves for food, water, air, light— everything that enables organisms to survive and reproduce. The second observation is that offspring differ slightly from their parents and from each other in heritable ways. This is variation. Darwin held that heritable variation consisting of slight differences between individuals, rather than larger mutations, was the source of evolutionary change.

Under conditions of superfecundity and a struggle for existence, according to Darwin, any variation that is useful to an individual tends to be preserved. Individuals that possess some slight advantage in securing means of subsistence and surviving to reproduce tend to leave more offspring than others. These offspring, in turn, possess the same advantage by inheritance. As generation follows generation the inherited advantages accumulate. Environments change; new variations prove advantageous; old variations prove detrimental. Organisms begin to show a great divergence of characteristics. Some become specialized to new habitats, others become extinct. Finally, separate interbreeding populations appear. These are new species. They have originated naturally, by the selective preservation of useful variations in changing environments.

Darwin called this process "natural selection" because he found it analogous to the "artificial selection" practiced by animal breeders. But in natural selection, according to Darwin, "Nature" is the selector, a set of laws ordained by God. All the

◀ **The first serious suggestions** about animal species being the result of development (rather than of specific creation) issued from the 18th-century intellectual ferment known as the Enlightenment. This emphasized the use of reason to scrutinize existing ideas. A typical freethinker was the physician Erasmus Darwin FAR LEFT. Because some animals metamorphose (eg tadpole into frog) he proposed that species are able to change and adapt to their environment. Jean-Baptiste de Lamarck LEFT CENTER explained change by suggesting that characteristics acquired by an animal could be transmitted to its offspring.

Etienne Geoffroy Saint-Hilaire LEFT ABOVE investigated the forms and developments of embryos and the similarities between different animals. He argued that individual species were variants developed from a single plan. Studies of "monstrous births" provided evidence of how change can occur.

◀ **A leading naturalist opponent** of incipient evolutionary thought was Georges Cuvier. He divided the animal kingdom into four types, thus opposing Geoffroy's concept of the unity of all living species. He insisted that each species was created for a specific purpose.

phenomena of life and mind have evolved progressively but purposelessly, except perhaps humankind. The "most important, but not the exclusive, means of modification" has been natural selection.

Pre-Darwinian Natural History

At the beginning of the 19th century very few naturalists believed that living things had evolved. Most of these naturalists were in France, where evolutionary ideas of "transformism" had become associated with materialism and revolution. Jean-Baptiste de Lamarck (1744–1829) devised a theory of organic change that was thoroughly materialistic. Living organisms are spontaneously generated from dead matter. Animals develop by their innate "power of life" along a linear scale of increasing complexity. They respond to changing environments by consciously altering their habits, and this causes structural changes that are passed on to their offspring. Eventually human beings are formed.

Lamarck produced little evidence for his theory, while much could be said against it. Geologists had only just begun to allow for the immense periods of time that it required. Paleontologists such as Georges Cuvier (1769–1832) had begun reconstructing fossil animals that Lamarck denied were extinct. Cuvier had conservative religious and political motives. He undermined Lamarck's concept of a linear progression of animals up to humankind by dividing the animal kingdom into four separate

"branches"—mollusks, radiates (eg starfishes), articulates (eg insects), and vertebrates—each with its unique organization. Animals, according to Cuvier, are perfectly adapted to their environments. Once out of adjustment, they do not evolve but become extinct.

An alternative to this view was being developed in Germany by philosophers such as Johann von Goethe (1749–1832) and Lorenz Oken (1779–1851). They opposed materialistic biology by arguing that Nature is the manifestation of a single plan or "idea." This "Nature Philosophy" was taken up in France by Etienne Geoffroy Saint-Hilaire (1772–1844), who pointed out fundamental structural similarities, or "unity of type," throughout the animal kingdom. Vertebrates, for example, all seem to have the same basic skeletal pattern. He rejected Cuvier's four branches; he denied that there was any strict relationship between animal structure and environment. Animal structures, according to Geoffroy, exist to manifest the unity of type. They change because the environment disturbs embryonic development. New species originate as monstrous births. In 1830, on the eve of the July revolution, Geoffroy and Cuvier debated publicly in Paris. The liberal régime of Louis-Philippe was installed, but the conservative Cuvier triumphed.

In Great Britain, only a few radicals such as Erasmus Darwin (1731–1802), the grandfather of Charles, sympathized with French transformism and Lamarck. Otherwise naturalists gave homage to Cuvier. Many were clergymen; almost everyone interpreted the world in the light of "natural theology." They believed that plants and animals were perfectly adapted to their environments because each species had been specially designed and created by God. Perfect and purposeful adaptations proved that the world was governed by a wise and beneficent Creator who had ordained social inequalities and forbidden revolution. Geologists such as the Revd William Buckland (1784–1856) and the Revd Adam Sedgwick (1785–1873) differed in their interpretation of the evidence for the Flood, although both gave a directional, "progressionist" account of earth-history in opposition to the cyclical, "uniformitarian" account of their colleague Charles Lyell (1797–1875). But all three naturalists agreed that species originated through acts of creative power inserting well-adapted forms into prepared environments. Perfect, purposeful adaptation by God was the sole explanation for the change. Evolution—and revolution—were precluded.

The Reform Bill was passed in 1832, bringing Britain one step closer to modern democracy. As agitation for social change continued over the next two decades many British naturalists adopted a new kind of explanation for changes in the history of life. They came to regard perfect adaptations of each individually created species to its environment as an inadequate account. For one thing, there seemed to be too many purposeless structures. Adaptation might prove to be less than perfect, but the naturalist's task was still to explain both the unity of type and the special adaptations within each of Cuvier's branches. Both together testified to God's omniscient design and furnished the basis for a new natural theology. By the time Charles Darwin began his researches on evolution, naturalists such as Richard Owen (1804–1892) and W. B. Carpenter (1813–1885) were arguing that the unity and diversity of species should be explained by general laws of nature.

Darwin's Theoretical Work

After returning from his voyage on HMS *Beagle* in 1836 (see p60), Darwin immersed himself in the researches of fellow naturalists. His changing views on the origin of species closely paralleled theirs. Steeped in the *Natural Theology* (1802) of Archdeacon Paley and in Lyell's *Principles of Geology* (1830–33), Darwin found it impossible to accept that all species had always been adapted to the same environment. He agreed with Lyell that the age of the earth was immense. The problem was that a species that had been created in a certain way and remained unchanged could not explain purposeless structures, unity of type, and the similarities between fossil and living inhabitants of a region where the environment had changed. "Look abroad," Darwin confided to a notebook, "study gradation, study unity of type, study geographical distribution, study relation of fossil with recent. The fabric falls!" The "fabric" Darwin had in mind was the whole fabric of traditional natural theology, with its social doctrines. But this was a dangerous thought, so Darwin kept it to himself. He had glimpsed a "grander" theology in which not only species but societies would change and improve according to laws ordained by God.

How then did species originate? Darwin had rejected created adaptation; he also denied the particular explanation offered by Owen and Carpenter, and the Nature Philosophy on which they drew. If unity of type and special adaptations were governed by a divine idea, as they claimed, then nothing followed. Such a theory led to no predictions that could be tested. And it demeaned the Creator to suppose that his "will" could be used as a basis for scientific statements. Darwin required a materialistic explanation for the unity and diversity of living things. He found his explanation—natural selection—by reading authors who had sought out the laws of change in society, among them the Revd Thomas Malthus (1766–1834). In his *Essay on the Principle of Population* (1798; 6th edn, 1826) Malthus had argued that superfecundity was good for people because it tended to improve their character. The threat of a geometrically increasing population (in the ratio of 1:2:4:8 . . .) with only an arithmetically increasing food supply (in the ratio of 1:2:3:4 . . .) would galvanize human energies and bring about a "gradual and progressive improvement" in society. In 1838 Darwin applied this principle to the rest of the living world. Superfecundity was the "force" that would "sort out structure and adapt it to changes" in the environment.

For 20 years Darwin concealed his views on evolution. Like Lamarck, he had devised a materialistic theory of evolution. Although his theory was deeply indebted to natural theology, it, too, would have appeared revolutionary because it applied to human beings. After reading Malthus, Darwin refined his concept of variation and began to question whether nature was a harmonious system after all. He set out to discover whether natural selection could explain recent generalizations (eg "unity of type") and solve current problems (eg the geographical distribution of species) in the various branches of natural history. In the mid-1850s he made the last major revision to his theory. He saw that species tend to diversify even in a stable environment, because natural selection favors increased specialization wherever there is competition. This was the "division of labor" of Victorian political economists. It became the "principle of divergence" in Darwin's first major public statement of his theory, a 500-page "abstract" entitled *On the*

Alfred Russel Wallace: Darwin's Co-discoverer

The reason why scientists today do not refer to "Wallacean evolution" is that a magnanimous young naturalist declined to engage in a nasty dispute over priority.

The career of Alfred Russel Wallace (1823–1913) was about as unlike Darwin's as one may care to imagine. He was a Welshman who left school at 13 and never went to university. He worked as a surveyor and a schoolmaster before sailing for the Amazon jungles in 1848 as a professional collector of natural history specimens; he hoped to solve the problem of the origin of species. His ambition to make the trip had been raised by Darwin's *Journal of Researches*, but his inspiration to find the "law" of evolution had come directly from an anonymous book by an amateur, *Vestiges of the Natural History of Creation*. The expedition ended in disaster. On his way home the

ship burned and sank, which destroyed Wallace's collections, his notes and sketches, and most of his journal.

Wallace set out again two years later, in 1854, for the Malay Archipelago. There in February 1858, while pondering the differences between "savage" and "civilized" people, he remembered the argument of Thomas Malthus' *Essay on the Principle of Population*, which

had been published in 1798.

He applied the argument to animals and plants, as Darwin had, and arrived independently at the theory of natural selection. When his manuscript, "On the tendency of varieties to depart indefinitely from the original type," reached a thunderstruck Darwin, a "delicate arrangement" was reached. Wallace received credit for a clear statement of the theory in a communication by Darwin's friends, Lyell and Hooker, to the Linnaean Society of London. This consisted of Wallace's paper, preceded by extracts from unpublished manuscripts that established Darwin's priority.

When Wallace returned to England in 1862 he remained the perfect gentleman. His relationship with Darwin cooled for other reasons. During the 1860s, while Darwin was admitting causes of evolution other than natural selection,

and explaining the evolution of humankind, Wallace did the reverse. He concluded that natural selection was all-powerful *except* in the case of the higher human faculties. The reasons for this divergence were complex. Wallace never quite fitted into professional circles; his views of human nature and society were unorthodox. He believed in spiritualism and the nationalization of land. He saw himself as a socialist. In 1907 he published a book entitled *Is Mars Habitable?*

But the co-discoverers of natural selection stayed on friendly terms. In 1881 Darwin used his influence to obtain a civil pension for Wallace. A year later Wallace bore Darwin's coffin in Westminster Abbey. In 1889 he published a masterful exposition of the theory of natural selection. The title he gave to this work was: *Darwinism*.

◄ **HMS *Beagle*,** a government-sponsored survey ship, in the Straits of Magellan in 1834. In 1831 Charles Darwin was recruited for the *Beagle*'s voyage to South America with the job of "collecting, observing, and noting, anything worthy to be noted in natural history." At that time he was particularly interested in geology. In his *Autobiography* he wrote: "The Voyage of the *Beagle* has been by far the most important event in my life, and has determined my whole career." His observations convinced him of the mutability of species. Within two years of the end of the voyage in 1836 he had hit on an explanation for this.

► **A missing link.** An essential part of Darwin's theory of natural selection came from the state of nature envisaged by the Revd Thomas Malthus ABOVE. Malthus pointed out that natural populations increased faster than the sources that sustain them. This causes competition and the elimination of the weak.

► **Charles Darwin in 1854,** four years before he began to write *On the Origin of Species*.

Origin of Species by Means of Natural Selection, or the Preservation of Favoured Races in the Struggle for Life (1859).

Post-Darwinian Controversies

Darwin focused a revolution in natural history, but it was not all his own doing. His theoretical work was shaped and informed by concepts of nature and society that were widely held in early-Victorian times. Before the *Origin of Species* appeared, nonscientists such as Robert Chambers (1802–71) and Herbert Spencer (1820–1903) had published theories of evolution by natural law. Chambers' anonymous *Vestiges of the Natural History of Creation* (1844) belonged to the tradition of Owen and Geoffroy. Spencer's essays leading to his *System of Synthetic Philosophy* (1860–96) developed a Lamarckian theory of inevitable progress, although they also expressed many of Darwin's own assumptions. These writings had accustomed the middle-class public to the idea of evolution, but the *Origin of Species* was specially controversial. It came from a respected naturalist; it contained a powerful argument that shattered a gentleman's agreement among Darwin's professional colleagues. In 1859 Darwin opened a Pandora's box of human problems when he theorized about origins and change. Old friends such as Sedgwick and Lyell grieved to tell him so.

Darwin's *Descent of Man, and Selection in relation to Sex* was published in 1871. The *Times* suggested ominously that the book had appeared as Prussia occupied Alsace and French troops besieged Communists in Paris. But the opponents of natural selection had long since spotted sinister implications in the struggle for existence. Not only Sedgwick and Lyell but leading churchmen such as Samuel Wilberforce (1805–73) and Conservative politicians such as Benjamin Disraeli (1804–81) had immediately deplored Darwin's theory of "the pithecoid origin of man." Natural selection placed human morals and society in an animal world that was undergoing continuous competitive change. To them this was not acceptable: man was above this.

The issues confronting Darwin's opponents in the 1860s, like their alliances, were complex. Human origins lay at the center of the debate but did not circumscribe it. Owen felt slighted that his idealistic quest for a "continually operating secondary

◄ **The spread of Darwinian evolution.** Darwin's theory of natural selection was attacked mainly for philosophical and theological reasons. Among leading scientists, however, it quickly gained ground. Joseph Hooker TOP, eminent botanist, showed that Darwinian theory could be applied to botany. In America the leading botanist Asa Gray MIDDLE published articles explaining and supporting natural selection (collected in *Darwiniana*, 1876), though as a Christian he also maintained that variation in species was partly of divine origin. Others equated Darwin's theory with their own. Herbert Spencer BELOW, an ardent believer in social progress, absorbed natural selection into his sociology, devising the phrase "survival of the fittest."

◄ **Darwin and friend,** LEFT, from the *London Sketch Book*, 1874.

► **Sexual selection:** heads of the South American catfish *Plecostomus barbatus*; male above, female below. Darwin explained: "When the sexes differ in external appearance, it is, with rare exceptions, the male which has been the more modified; for, generally, the female retains a closer resemblance to the very young of her own species. . . . The cause of this seems to lie in the males of almost all animals having stronger passions than the females. Hence it is the males that fight together and sedulously display their charms before the females; and the victors transmit their superiority to their male offspring."

creational law" had been forestalled. He coached Wilberforce's attack and inspired younger anti-Darwinian evolutionists such as the Duke of Argyll (1823–1900), a Liberal statesman, and St George Mivart (1827–1900), a Roman Catholic zoologist. Meanwhile some of Darwin's assumptions were being formidably disputed by physical scientists with less obvious axes to grind. William Thomson (1824–1907), later Lord Kelvin, argued from the laws of thermodynamics that geological time had been insufficient for natural selection to be the mechanism of evolution. Fleeming Jenkin (1833–85), an engineer, argued that neither slight variations nor larger mutations could be accumulated by natural selection to form new species on Darwin's assumption that offspring inherit a blend of parental characteristics. Objections like these were grist to the mill of innumerable small-time critics who had overtly religious and ideological objections to Darwin's theory.

For his part Darwin could count on support from a rapidly

The Descent of Women: Sexism in Evolution

Most evolutionists in the 19th century believed that human societies had evolved according to laws that applied throughout the organic world. They attempted, therefore, to explain existing social phenomena by known natural laws. One such phenomenon was the role and character of women.

Evolution was often supposed to be a process analogous to embryonic development. According to this so-called *ontogenetic analogy*, the human embryo recapitulates its animal ancestry, and the growing male child exhibits in sequence the physical and mental characteristics of subordinate social groups, including lower races and women. These groups have experienced "arrested development": that is, the diminution of organic energy, or the premature diversion of that energy into physical rather than mental growth. In women the energy diverted from mental development is required by the complex structure and function of their reproductive organs. Evolution therefore was held to explain the supposed intellectual superiority of men.

Evolution was also often supposed to be a process that distributed energy in society by analogy with the distribution of energy in individual organisms. According to this *organic analogy*, individual societies have definite amounts of energy, and the more advanced societies have greater amounts of highly differentiated energy. Such societies have evolved complex "structures" that are well adapted for performing vital "functions," just as the higher animals have. In the most advanced groups of the most advanced societies, the reproductive function is performed in the monogamous nuclear family, just as in the higher female organisms reproduction takes place in the womb. Evolution therefore has not only equipped women to reproduce the species; it has adapted society to provide the ideal place for them to do this.

Darwin subscribed to these myths about women with a theory that he perfected in the 1860s, when the "woman question" was being debated in Victorian Britain. He published it in the second and third parts of the *Descent of Man*. According to this theory of *sexual selection*, females throughout the animal kingdom have selected the most impressive of the males that competed to reproduce with them. This selection, however, was purely passive and aesthetic. Male aggressiveness made females coy, impressionable and capricious. Women have retained these characteristics through inheritance, together with their arrested intellectual powers and maternal instincts.

Wallace flatly disagreed with

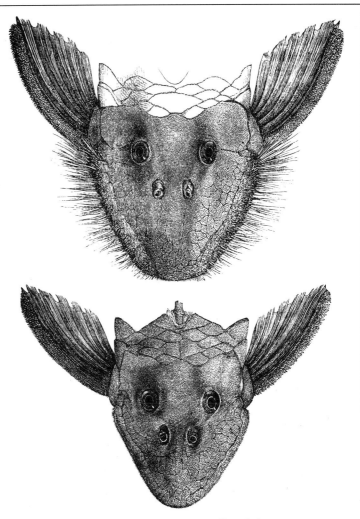

Darwin. Female animals had not been selective agents; natural selection was supreme. But human societies are no longer subject to natural selection. The future, according to Wallace, belongs to independent, educated women, whose rational selection of the worthiest men in marriage is the foundation of social progress.

increasing number of naturalists and liberal intellectuals. In general, adherence to natural selection was strongest among his friends. Lyell never went "the whole orang," as he put it. T. H. Huxley (1825–95) never defended natural selection with much enthusiasm. But neither was as close to Darwin as the botanists J. D. Hooker (1817–1911) in Britain and Asa Gray (1810–88) in the USA. They became the foremost senior Darwinians in the English-speaking world. By the 1880s Darwin had personally inspired a circle of younger Darwinians in Britain, including John Lubbock (1834–1913), E. Ray Lankester (1847–1929), George Romanes (1848–94) and E. B. Poulton (1856–1943). In Germany Ernst Haeckel (1834–1919) popularized a form of Darwinism in the guise of the old Nature Philosophy.

In five successive revisions of the *Origin of Species* (1860–72) and in the *Descent of Man* Darwin replied to the critics of his theory. He conceded that the direct action of the environment and the habitual use or disuse of parts were independent causes of evolutionary change where natural selection had been found inadequate. This concession encouraged critics such as Samuel Butler (1835–1902), George Henslow (1835–1925), and the American "Neo-Lamarckians," to claim that Lamarck and Geoffroy had been right after all. Natural selection was not *the* mechanism of evolution. "Neo-Darwinians" such as Weismann leapt to the defense, but his famous war of words with Spencer in the 1890s indicated how largely, at the time, theories were assessed on the basis of the logical improbability of their rivals rather than by means of empirical evidence. Spencer commented, "A right answer to the question of whether acquired characters are or are not inherited [that is, by the offspring], underlies right beliefs, not only in Biology and Psychology, but also in Education, Ethics, and Politics." In this single remark he clarified once and for all the profound issues at stake. (See also "Controversies of Evolution.")   JRM

Charles Darwin

The naturalist in society

The society into which Charles Darwin was born on 12 February 1809 was noted for liberality in politics and religion. Both Charles' grandfather, Erasmus Darwin (1731–1802), and his father, Robert Darwin (1766–1848), were physicians, Fellows of the Royal Society, and freethinkers. Susannah Darwin (1765–1817), his mother, was, like her Wedgwood relatives, a Unitarian (ie a believer in God as one person, not three-in-one). The Darwins and the Wedgwoods lived comfortably in the West Midlands of England, not far from the commercial centers that sparked off the Industrial Revolution. They and their friends believed in progress and social reform. They admired the ideals of the French Revolution. They were the avant-garde of English middle-class Dissent.

Nevertheless, when Charles—having studied at Edinburgh University—had failed to take up his father's profession, Dr Darwin sent him to study at Cambridge with a view towards a career in the Church. Already the young man had shown a flair for natural history. A comfortable country parish was not beyond Dr Darwin's means; there his son might disport himself as he pleased while pursuing a respectable vocation. The late-Georgian Church of England was very very broad.

Charles spent his years at Cambridge, from 1828 to 1831, in the company of ordinands and reverend professors. Many were naturalists; one of them was his tutor, the professor of botany and assistant curate of St Mary-the-Less, John Stevens Henslow (1796–1861). Henslow had his students' best interests at heart. When an offer came for a naturalist to join HMS *Beagle* on a surveying voyage round the world, he himself declined it reluctantly and offered the chance to Charles. Unlike Henslow and other candidates, Charles was neither married, nor beneficed, nor as yet ordained. His uncle, Josiah Wedgwood, convinced Dr Darwin that the voyage would not damage his son's future reputation as a clergyman.

On the *Beagle* voyage Charles achieved intellectual emancipation. He learned to think for himself. While Parliament passed the first great piece of middle-class legislation in 1832, Charles was undergoing a Reform Act of the mind. At Cambridge the works of the theologian Paley had "charmed and convinced" him, and the writings of Alexander von Humboldt and John Herschel had fired his enthusiasm for natural history. Now, through bouts of seasickness and debates with the Tory captain, Robert FitzRoy (1805–65), he began to look at God's creation in a new way. He was awestruck by a tropical forest, by the power of an earthquake, by the malleability of human beings. There was little to choose between Fuegian indians and colonial Gaucho "savages"; it made him wonder whether human nature had not evolved. The superfecundity of living things made him wonder why so much beauty had been created for "such little purpose." The minute adaptations of animals for life on the separate islands of an archipelago shook his assumption that the Creator had specially designed each species for its environment—this was the doctrine of Paley and Lyell. Charles pored through the first two volumes of Lyell's *Principles* when he arrived back at Falmouth in October 1836. By now Paley's *Natural Theology* had begun to look antiquated.

From 1836 to 1842 Charles sorted through the rich debris of the voyage. There were zoological specimens to be described, geological theories to spin, and his *Journal of Researches* (1839)

to publish. Henslow in Cambridge continued to advise him, while Lyell, a new mentor, paved his way into scientific London. Charles chose to live in the city near his unmarried elder brother, Erasmus. Publicly it was the period of his greatest prominence. Privately these were his most creative years. Almost every subject on which he later theorized and published was canvased in a series of notebooks that he kept until *c.* 1840. There he confided his ideas about natural selection. Charles conceived the theory in the autumn of 1838, just after he and his father had resolved the problem of his career. Dr Darwin approved his marriage to Emma Wedgwood (1808–96), a first cousin, and promised to keep on subsidizing his research.

Charles and Emma married early in 1839 and settled in London for the time being. The idea of ordination was forgotten as Charles became obsessed with his theories. But on consideration he decided that a clerical lifestyle, like Henslow's, would suit him best. Since pondering evolutionary ideas he had felt increasingly ill. Besides, his beliefs were unorthodox and potentially dangerous, even though he still believed in God. In 1842 Charles left Lyell and Erasmus behind and retired with Emma to the village of Downe in Kent, a safe and comfortable 25km (16mi) from the metropolis. Together they populated the old parsonage, Down House, with ten children, seven of whom survived. There were servants and gardens and a greenhouse for experiments. The family attended the parish church and

◀ **Darwin's seat of learning** FAR LEFT, Down House at Downe in Kent, about 25km (16mi) from London; an exterior view and the study. Darwin lived and worked here for 40 years (1842–82).

◀ **Late in life:** Darwin photographed shortly before his death. A constant support and unsung heroine of his achievement was his wife INSET.

▼ **One of "Punch's Fancy Portraits,"** Darwin caricatured in 1881. The caption read: "In his *Descent of Man* he brought his own species down as low as possible – i.e. to 'A Hairy Quadruped' . . . He has lately been turning his attention to the 'Politic Worm'." (At that time of his life Darwin had just published his study of earthworms.)

the children were baptized there. Charles invested in railroads and voted Liberal. He became a magistrate, the treasurer of the local Friendly Society, and a pillar of the parish.

The latter half of Charles' life contrasts sharply with the former. These were 40 years of unremitting routine, apart from sojourns at health spas, a few holiday trips, and occasional visits to London. Charles spent the time quarrying his old notebooks and scientific periodicals for facts to support his theories. He solicited data from scores of naturalists around the world. He spent about eight years dissecting "vile molluscous creatures" (barnacles) in order to study the problems of classification. In 1871 he began to abstain from controversy; his health improved. So he took up dissecting plants.

Meanwhile there was an unceasing flow of monographs from Down House. Charles remarked in 1876 that his mind seemed to be "a kind of machine for grinding out general laws from large collections of facts." Certainly 16 books and 152 other publications, even from a gentleman of independent means, is evidence of more than prolonged hard work. History, however, must not only remember his intellectual fecundity, but the strength and forbearance of Emma. The renegade ordinand died of a heart attack on 19 April 1882 and was buried in Westminster Abbey. She, the potter's daughter, survived him by 14 years and was laid to rest at Downe.　　　　JRM

Bringing Nature into Order

Principles of animal classification

The fundamental unit of animal classification is the species. Its members are so alike genetically that, not only are they very similar in most of their physical and other characters, but they should also be capable of interbreeding to produce fertile offspring.

The interbreeding test, of course, cannot be applied to fossil species, which because of their continual evolution vary through time as well as through space. It is therefore necessary to demarcate fossil species in an arbitrary manner, either by choosing certain characters as diagnostic of each species, or by delimiting them on the first appearance of certain evolutionary novelties; or, alternatively, by comparing all other specimens with the material first described under that name (primary type-material), the absolute standard of the species. The type concept is far more important in paleontology than in modern zoology.

Species themselves are classified further into larger, more comprehensive categories (higher taxa), but it cannot be emphasized too strongly that all such classifications are man-made, each to suit a particular purpose, and that they are therefore subjective. There can be no such thing as a "correct" classification. For a specified purpose, however, one classification may be preferred to another.

Systems of Classification

For scientific purposes species are traditionally classified hierarchically (that is, into nested sets) according to how alike they are, the degree of their similarity being judged mainly on physical characteristics. In the early days of systematic biology (from the middle of the 18th century), before the true nature and significance of fossils had become apparent, each group of similar organisms was thought to stem from an archetypal plan in the mind of the Creator; thus the hierarchical arrangement of the living taxa would actually be based on nothing more than a subjective appraisal of overall similarity, without reference to any objective scientific criterion. This approach seemed to work very well in most cases, for the majority of living forms fall easily into taxa that are clearly characterized by sets of diagnostic features. For example, no reasonably intelligent person would find it difficult to distinguish a bird from a "nonbird," for all birds and only birds possess feathers and a furcula (wishbone).

During the 19th century paleontology developed rapidly. The idea of organic evolution gradually became accepted by most systematists; they adopted the practice of using in classification only those similarities that they believed to be due to common ancestry, so that the hierarchical arrangement of the categories would reflect to some extent the supposed pattern of branching of the phylogenetic tree (the "family tree" of all groups of animals) and would in no way contradict it. One general principle of such a classification is that the more recent the common ancestry of two groups of organisms is supposed to be, the closer their "relationship" and the more closely together should they be classified; on the other hand, a rapid burst of evolution leading to the origin of a new and very different group may make it desirable for practical purposes to segregate that group into a new category, separate from the forms which are closest to it in the phylogenetic tree. Thus the crocodiles share a more recent common ancestry with the birds than they do with other groups of extant reptiles; yet the crocodiles are classified with those others in the class Reptilia, while the birds, in view of their having made such a huge evolutionary leap, are placed in a class of their own (Aves).

One consequence of this approach is that some of the higher taxa do not include all their own descendants; that is, they are paraphyletic. The Reptilia are a paraphyletic taxon, for they gave rise to the Aves (birds) but do not include them. The Aves and the Mammalia, by contrast, are holophyletic taxa, for they have given rise to nothing except other birds and mammals respectively.

The incorporation of fossil animals into zoological studies increased the understanding of the subject; yet, at the same time, it made classification more difficult. There were three reasons for this. First, many of the newly discovered fossil species did not fit comfortably into the established higher categories, for they possessed some of the diagnostic characters supposedly typical of one category and some of another. Secondly, many fossils (especially vertebrates) are unique or few in number, often very incomplete or mere fragments, and almost invariably they show only skeletal structures. Thirdly, the whole matter is rendered far more complex by the addition of another dimension—that of time.

The "orthodox" or Simpsonian classification just described is what most systematists and most other biologists still use today. It represents a sensible compromise between conflicting objectives, none of which, in any case, is fully attainable.

Within recent years, however, new approaches to classification have appeared, over which a great deal of controversy has arisen. Two of these are together referred to as cladistics. Such approaches are typified by, first, a more rigorous, formalized method of character analysis that produces a dichotomously branching cladogram: that is, a tree diagram that branches into two at each node, according to whether a given character is present or not; secondly, an hierarchical classification that corresponds in every detail with that cladogram; and thirdly, a requirement that all taxa should consist of complete clades, ie holophyletic groups. Paraphyletic taxa are forbidden.

The two types of cladistics are diametrically opposed in one important fundamental. In the original type (Hennigian or phylogenetic systematics) the only criterion of the classification is the phylogenetic tree, which the cladogram is assumed to represent. Out of this has evolved the contrasting "transformed" cladistics (natural order systematics) in which the cladogram has no phylogenetic significance whatever; it merely represents the "natural order," the most parsimonious arrangement of the species (ie that which is supported by the greatest number of characters). Such an arrangement may be interpreted, but only if so desired, as the result of evolution.

Another type of classification employed by some systematists since the 1960s is phenetics; when it is quantified more precisely it is called numerical taxonomy. The species are grouped together on the basis of their overall similarity, using as many characters as possible and according them all equal importance; members of a group need not all share the same unique diagnostic features, and their phylogenetic history is ignored. Phenetics is used more for plants than for animals.   AJC

SCHEMES OF CLASSIFICATION

A series of diagrams and charts illustrating some of the differing approaches to classification. The example taken to show the systems is that of the main groups of terrestrial vertebrates, ie amphibia, reptiles, birds and mammals, although some of the techniques are not normally applied to this level of taxa.

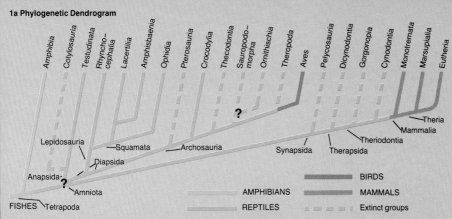

1a Phylogenetic Dendrogram

Amphibia · Cotylosauria · Testudinata · Rhyncho-cephalia · Lacertilia · Amphisbaenia · Ophidia · Pterosauria · Crocodylia · Thecodontia · Sauropodo-morpha · Ornithischia · Theropoda · Aves · Pelycosauria · Dicynodontia · Gorgonopia · Cynodontia · Monotremata · Marsupialia · Eutheria

Lepidosauria · Squamata · Archosauria · Theria · Mammalia · Theriodontia · Synapsida · Therapsida

Diapsida · Anapsida? · Amniota

FISHES · Tetrapoda

? (Aves)

BIRDS
AMPHIBIANS
MAMMALS
REPTILES
Extinct groups

1b "Orthodox" (Simpsonian) Classification

Amphibia
Reptilia
 Anapsida
  †Cotylosauria
  Testudinata
 Diapsida
  Lepidosauria
   Rhynchocephalia
   Squamata
    Lacertilia
    Amphisbaenia
    Ophidia
  Archosauria
   †Thecodontia
   †Pterosauria
   Crocodylia
   †Sauropodomorpha
   Ornithischia
   †Theropoda
 †Synapsida
  †Pelycosauria
  †Therapsida
   †Dicynodontia
   †Theriodontia
    †Gorgonopia
    †Cynodontia
Aves
Mammalia
 Monotremata
 Theria
  Marsupialia
  Eutheria

†Extinct groups

Paraphyletic groups

1c "Phylogenetic" (Hennigian) Classification

Tetrapoda
Amniota
 Anapsida
  †Cotylosauria
  Testudinata
 Diapsida
  Lepidosauria
   Rhynchocephalia
   Squamata
    Unnamed taxon
     Lacertilia
     Amphisbaenia
    Ophidia
  Archosauria
   †Pterosauria
   Crocodylia
   †Thecodontia
   †Sauropodomorpha
   †Ornithischia
   †Theropoda
   †Coelurosauria
   Aves
 Synapsida
  †Pelycosauria
  Therapsida
   †Dicynodontia
   Theriodontia
    †Gorgonopia
    Cynodontia
     Mammalia etc

▲▶ **Old and modern schemes.** (1a) A phylogenetic tree from which both (1b) "orthodox" (Simpsonian) and (1c) "phylogenetic" (Hennigian) classifications may be derived. Phylogenetic cladists make less use of evidence other than character distribution, eg from stratigraphy and geography, in the construction of the dendrogram than do orthodox systematists. Note that in the "Orthodox" (Simpsonian) classification, groups such as amphibia and reptiles do not include all their descendants, ie birds, mammals; that is amphibia and reptiles are paraphyletic. In the "Phylogenetic" (Hennigian) classification such paraphyletic groups are not allowed and do not have a place in the classification scheme.

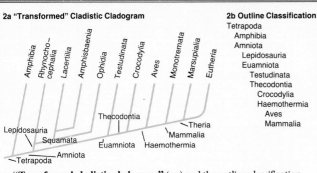

2a "Transformed" Cladistic Cladogram

Amphibia · Rhyncho-cephalia · Lacertilia · Amphisbaenia · Ophidia · Testudinata · Crocodylia · Aves · Monotremata · Marsupialia · Eutheria

Lepidosauria · Squamata · Thecodontia · Theria · Mammalia · Haemothermia · Euamniota · Amniota · Tetrapoda

2b Outline Classification

Tetrapoda
 Amphibia
 Amniota
  Lepidosauria
  Euamniota
   Testudinata
   Thecodontia
    Crocodylia
    Haemothermia
     Aves
     Mammalia

▲ **"Transformed cladistic cladogram"** (2a) and the outline classification (2b) derived from it. Some extreme "transformed cladists" insist that the cladogram must be based upon living (extant) forms only; and that fossil forms, if considered at all, should be inserted into that cladogram later; but only if and where possible such that they do not modify its fundamental structure. This cladogram shows only living forms.

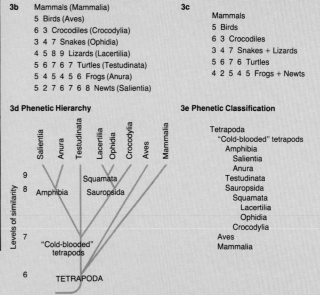

3b

Mammals (Mammalia)
 5 Birds (Aves)
 6 3 Crocodiles (Crocodylia)
 3 4 7 Snakes (Ophidia)
 4 5 8 9 Lizards (Lacertilia)
 5 6 7 6 7 Turtles (Testudinata)
 5 4 5 4 5 6 Frogs (Anura)
 5 2 7 6 7 6 8 Newts (Salientia)

3c

Mammals
 5 Birds
 6 3 Crocodiles
 3 4 7 Snakes + Lizards
 5 6 7 6 Turtles
 4 2 5 4 5 Frogs + Newts

3d Phenetic Hierarchy

Levels of similarity

Salientia · Anura · Testudinata · Lacertilia · Ophidia · Crocodylia · Aves · Mammalia

9
8 Amphibia · Squamata · Sauropsida
7 "Cold-blooded" tetrapods
6 TETRAPODA

3e Phenetic Classification

Tetrapoda
 "Cold-blooded" tetrapods
  Amphibia
   Salientia
    Anura
  Testudinata
  Sauropsida
   Squamata
    Lacertilia
    Ophidia
   Crocodylia
 Aves
 Mammalia

◀▲ **Phenetic classification**—stages of preparation. (3a) A table is constructed whereby for the selected animal groups the presence or absence of a certain character is keyed. (3b) From this a similarity table is constructed which indicates the number of similarities there are between each pair of animal groups. (3c) Those pairs with high similarity (ie 8 or 9 characters in common) are combined and the similarity table redrawn. (3d) A phenetic hierarchy chart can then be drawn from which (3e) a classification system is derived. This system of classification would not normally be applied to this higher group of taxa, nor to fossils. It is mainly used for recently evolved groups which show great diversity at a low taxonomic level (eg family, genus) as in plants and insects.

3a Phenetic Similarity Table

	Newts	Frogs	Turtles	Lizards	Snakes	Crocodiles	Birds	Mammals
Endothermic ("warm-blooded")								
Uricotelic nitrogen excretion								
Secondary palate								
Two occipital condyles								
Loss of teeth, horny beak present								
Reduction or loss of tail								
Four limbs								
Sigmoid locomotion								
"Fully-improved" stance								
Amnion present								

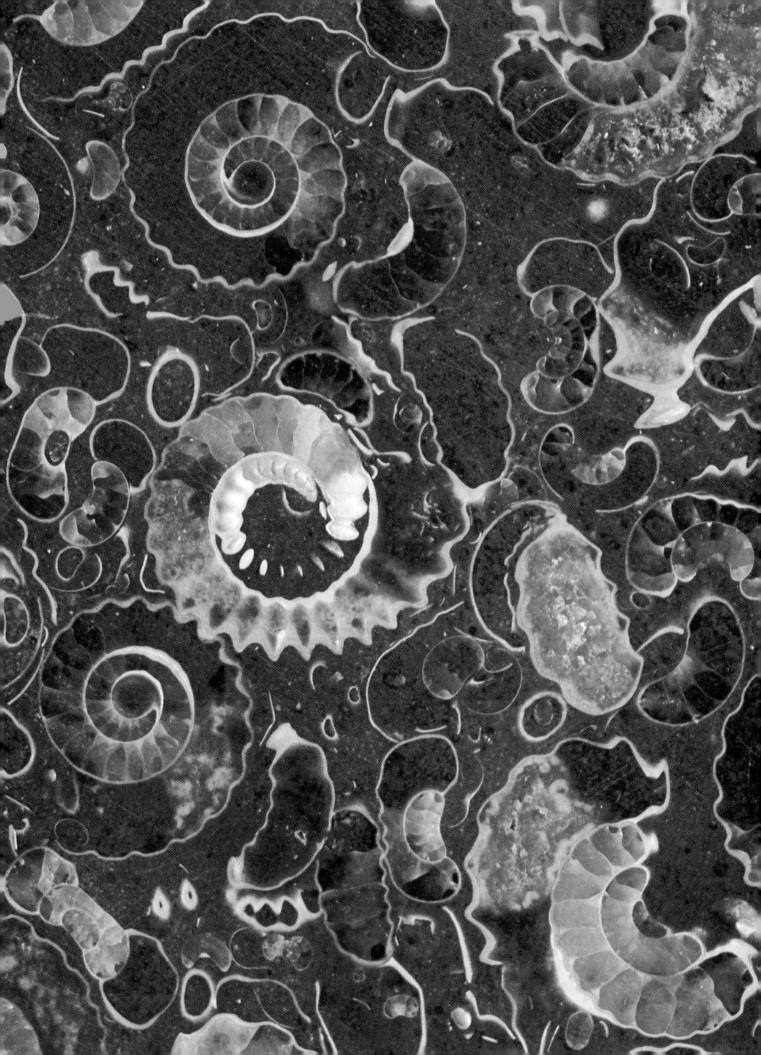

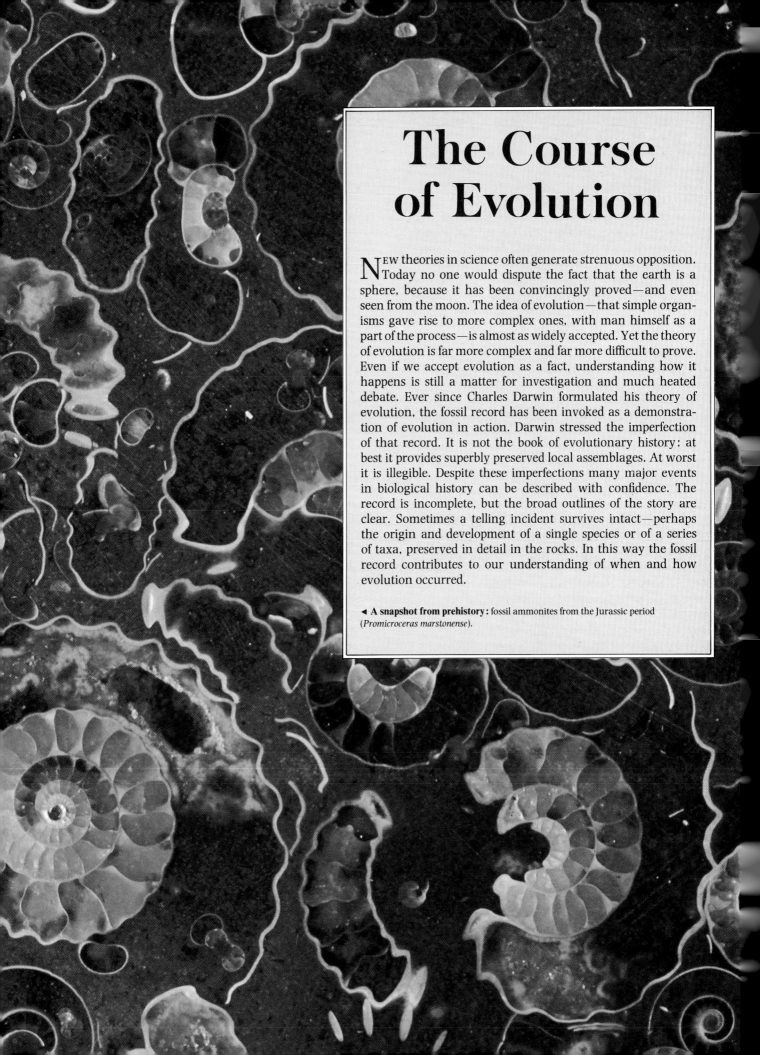

The Course of Evolution

NEW theories in science often generate strenuous opposition. Today no one would dispute the fact that the earth is a sphere, because it has been convincingly proved—and even seen from the moon. The idea of evolution—that simple organisms gave rise to more complex ones, with man himself as a part of the process—is almost as widely accepted. Yet the theory of evolution is far more complex and far more difficult to prove. Even if we accept evolution as a fact, understanding how it happens is still a matter for investigation and much heated debate. Ever since Charles Darwin formulated his theory of evolution, the fossil record has been invoked as a demonstration of evolution in action. Darwin stressed the imperfection of that record. It is not the book of evolutionary history: at best it provides superbly preserved local assemblages. At worst it is illegible. Despite these imperfections many major events in biological history can be described with confidence. The record is incomplete, but the broad outlines of the story are clear. Sometimes a telling incident survives intact—perhaps the origin and development of a single species or of a series of taxa, preserved in detail in the rocks. In this way the fossil record contributes to our understanding of when and how evolution occurred.

◄ **A snapshot from prehistory:** fossil ammonites from the Jurassic period (*Promicroceras marstonense*).

THE COURSE OF EVOLUTION

*How evolution proceeded. . . The evidence for when life
began on earth. . . The evolution of many-celled
animals. . . The conquest of land by plants and animals. . .
Subsequent evolution, including the conquest of the air. . .
How is evolution affected by the extinction of species?. . .
The dramatic effects of major extinctions. . . Why did the
dinosaurs become extinct while other groups survived?. . .
Regular periods of extinction in the sea. . . Life histories
and understanding evolution. . . Animal growth is never
uniform. . . Missing links: crucial fossils that are hard to
find. . . Missing links in man. . . The Piltdown hoax. . .
The early hominids. . . Lineages: evolutionary sequences
read out of rocks. . . Evolutionary trends. . . Living
fossils*

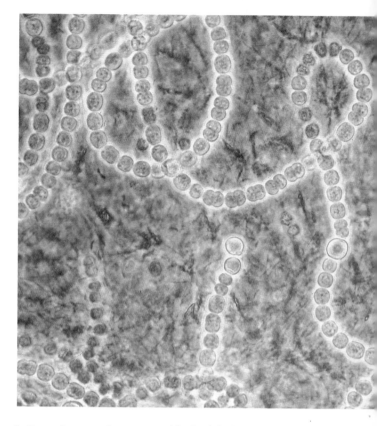

THE story of life's evolution, from simple to complex, from single-celled to many-celled, and from marine to terrestrial, is well known. But these major changes were not always achieved by a gradual progression. The biggest steps in evolution came when new thresholds were crossed, allowing organisms to proliferate in number and variety as never before. Soon after plants colonized the land a variety of animals were able to earn a living as herbivores. The success of these animals in turn provided a food source for terrestrial carnivores; and all such new life forms opened up further ecological opportunities, for example for parasites and scavengers. It might be compared with the effect of crossing new technological thresholds. Microelectronics, for instance, has generated a vast array of satellite industries, many of them only distantly related to the original concept. Crossing these evolutionary thresholds generated new designs and new patterns of life, and led to the conquest of entirely new habitats.

Some of the major events in the story of life have left evidence in the fossil record, but the most important of all—the origin of life itself—is not directly recorded. The earliest organisms were too insubstantial to leave any traces in the rocks, and these early rocks have themselves been heated, cooled, and subjected to pressure over millions of years, destroying any fossils they might have contained. Even so, it has been suggested that small carbon-rich spheres in very ancient rocks from Greenland are of biological origin. They have even been compared with similar spheres found in certain meteorites. The idea is that life "invaded" this planet from space. Few scientists are prepared to accept that these spheres *are* of biological origin. Most

believe that conditions suitable for life developed on the earth about 3,500 million years ago. Fossil remains of simple rod-shaped organisms similar to primitive bacteria are known from rocks about 3,000 million years old. Less than 1,000 million years later, algae capable of photosynthesis had evolved. Their remains are sometimes quite prolific in siliceous rocks deposited between 1,750 and 1,500 million years ago. These primitive plants were no more than simple threads, but they were vital to the whole subsequent history of life. Their photosynthesis enriched the atmosphere with oxygen—it may even have created the conditions in which more complex multicellular organisms could thrive.

The evolution of multicelled animals was one of the most important developments in the history of life. There is virtually no evidence for such organisms in rocks more than 1,000 million years old, yet by between 700 and 600 million years ago there were enough of them to leave a wide variety of fossils in many parts of the world. This is even more remarkable when you remember that animals without skeletons have a very low

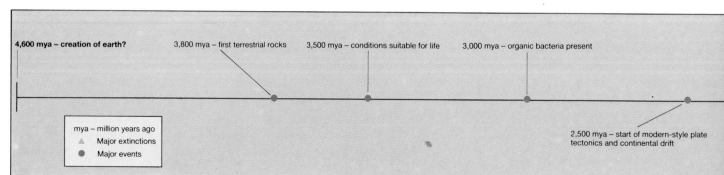

4,600 mya – creation of earth?

3,800 mya – first terrestrial rocks

3,500 mya – conditions suitable for life

3,000 mya – organic bacteria present

mya – million years ago
▲ Major extinctions
● Major events

2,500 mya – start of modern-style plate tectonics and continental drift

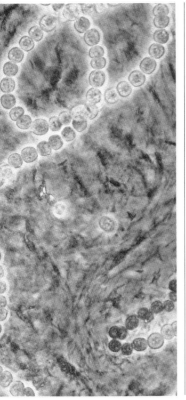

◄ **Chain gangs that filled the air:** strands of blue-green algae (genus *Nostoc*). Blue-green algae are a primitive cellular form of life (known as prokaryotes). Individuals lack the nuclei found in other cells and some other cell features. Such organisms made a vital contribution to the development of life, by putting oxygen into the atmosphere.

▼ **An evolutionary tale:** BELOW the major developments and extinctions of evolution set against a constant time scale.

▼ **Proliferating invertebrates.** While new forms of life evolved, many older forms were often being forced to adapt to new environments. Invertebrates, for example, developed significantly in the Ordovician period (505–438 million years ago). Basic body forms, such as the five-fold symmetry (pentamerism) seen in this Ordovician fossil brittle star (*Ophioderma egertoni*), remained the same but animals adapted in less spectacular ways to numerous different environments.

chance of being preserved as fossils at all. The seas must surely have been teeming with jellyfish and worms, as well as other animals that have no living counterparts. Compared with the long history of the plants that preceded them, their origin seems rather sudden, but their rarity as fossils makes this hard to prove.

The next major development was the evolution of hard skeletons, which are readily fossilized and make the record in the rocks a good deal easier to interpret. Many different kinds of organism seem to have developed skeletons during the same relatively short period. Varied fossil shells occur in rocks deposited over a few million years at the base of the Cambrian period, about 600 million years ago. These include a number that can be placed with confidence into groups of animals still living today. Mollusks and sponges were present, for example, and the commonly-fossilized trilobites belong to the arthropods, the great group of jointed-limbed animals that includes crabs and insects. Yet there are other, curious creatures that have no obvious living relatives. These are often described as evolutionary "experiments"—species that died out without producing descendants. No one has yet produced an entirely satisfactory explanation for the development of so many different kinds of animal with a hard skeleton, but the results are very clear: a "snowballing" process of expansion. New marine habitats became available to animals with hard parts, and it became possible for complex communities to evolve.

Life continued to proliferate in the sea. Before the end of the Cambrian period fossil remains indicate the presence of all the important animal phyla. By about 500 million years ago the marine ecosystem may have been complex. The next important stimulus to profound evolutionary change was the conquest of an entirely new habitat—the colonization of the land.

The first steps were unspectacular. During the Silurian period (438–408 million years ago) plants and animals began to colonize freshwater habitats. By the end of this period some plants had overcome the tremendous physical problems involved in growth without the support and protection of water. Creeping at first, then producing small shrubs, the plants created a new environment, rich in possibilities for animals to exploit. Fossils from the Devonian period (408–360 million years ago) show that relatives of the spiders and insects were quick to do so, while the first vertebrates had already made the difficult transition from fin to limb. Remains of these pioneer species are rare, but the rich fossil deposits of the Carboniferous period (360–286 million years ago) show just how successful they were. By this time a huge variety of terrestrial flora and fauna had

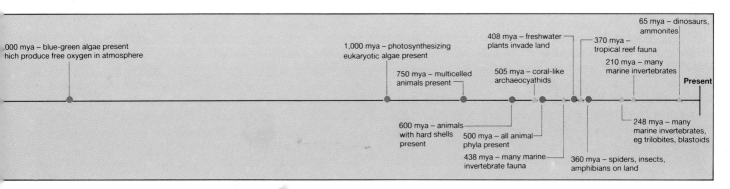

,000 mya – blue-green algae present hich produce free oxygen in atmosphere

1,000 mya – photosynthesizing eukaryotic algae present

750 mya – multicelled animals present

600 mya – animals with hard shells present

505 mya – coral-like archaeocyathids

500 mya – all animal phyla present

438 mya – many marine invertebrate fauna

408 mya – freshwater plants invade land

360 mya – spiders, insects, amphibians on land

370 mya – tropical reef fauna

210 mya – many marine invertebrates

248 mya – many marine invertebrates, eg trilobites, blastoids

65 mya – dinosaurs, ammonites

Present

evolved. Plants had acquired the stature of trees, there were giant, winged insect species, and some amphibians had grown almost as large as crocodiles. Seen against the vast span of geological time this was a relatively rapid change, accelerating with the impetus of its own success.

After the conquest of the land the fossil record gives ample evidence of the evolutionary changes that took place among both plants and animals. It shows the rise and decline of the giant reptiles, the emergence of mammals from small insectivores to their present dominance, and the spread of flowering plants after their origin in the Cretaceous period. As always, the appearance of one kind of organism allowed the success of others. The evolution of grass made possible the evolution of modern herbivores. The evolution of flowers allowed nectar-feeders such as butterflies to evolve, while butterflies and bees themselves encouraged the evolution of yet more wonderful and elaborate flowers.

The conquest of the last great ecological realm—the air— was more haphazard. It occurred several times among unrelated organisms, and each developed its own solution to the problem of flight. First, more than 300 million years ago, came the insects. They used delicate membranes drawn from their exterior skeletons to flit, and later to fly with the speed of a dragonfly. The pterodactyls, the flying reptiles of the Mesozoic era, became the greatest gliders the world has ever seen, using membranes stretched along their arms and one enormously elongated finger. The birds began as contemporaries of the flying reptiles, but outlasted them, developing superb control of powered, feathered flight. Lastly came the mammals, taking to the air at night while the birds were inactive. Bats, with their sophisticated navigation systems and aerial artistry, are in many ways the supremely specialized fliers. Several other kinds of animals, including lizards and marsupials, learned to glide. Some have been preserved as fossils. It seems that the apparently precarious threshold between the land and the air was less intimidating than the first, great boundary between the water and the land.

Extinction

Over the hundreds of millions of years that complex animals and plants have lived on earth, a vast number of species have become extinct. These extinctions are closely related to the origin of new species: extinction is a sign and opportunity for change. Today species are becoming extinct at an alarming rate, largely due to man's excessive impact on the environment. To see the historical perspective it is necessary to turn to the fossil record and ask to what extent extinctions are scattered and random, and to what extent they are concentrated at certain times or places. Perhaps times of extinction can be matched with other events that have left traces in the rocks. There may even have been real catastrophes, periods of crisis in the history of life.

Certain kinds of extinction are, in one sense, unimportant. When a species "splits" into two or more daughter species the ancestral species may, in time, become extinct—but it "lives on" in its descendants. This kind of extinction is probably fairly common, and is shown in the record of many fossil invertebrates. Rare species, such as those that evolved on isolated islands, are also likely to become extinct if their frail habitat is disturbed, but such species are hardly ever found as fossils.

At the other end of the scale is the extinction of a whole group of organisms, perhaps including thousands of species, which die out entirely, leaving no descendants. The dinosaurs are the best-known example, but there are many others: trilobites, a prolific group of arthropods that swarmed in the Paleozoic seas; ammonites, once so common that their fossils can sometimes be collected by the sackful; or planktonic graptolites, which disappeared during the early Devonian period after drifting successfully through the Lower Paleozoic oceans for 100 million years. The list of these major extinctions is a long one, and where several of them coincide it is reasonable to look for some major historical event as the cause. Some scientists have tried to explain all major extinctions as random events—rather like a long run of "bad luck" in a dice game—but if a whole range of organisms are all affected at the same time, "bad luck" seems a weak explanation.

The most notorious of these major extinctions took place at the end of the Cretaceous period, some 65 million years ago, when the dinosaurs disappeared. In terms of geological time, it seems to have happened very suddenly. The fall of these dominating giants has captured public imagination, but it is important to understand that many other organisms suffered extinction at the same time. The ammonites, a group with an even longer history than the dinosaurs, disappeared. So did the giant marine "lizards," the ichthyosaurs and plesiosaurs. Even the humble, single-celled foraminifera that floated in the ocean's surface waters were reduced to a fraction of their previous abundance. Obviously these facts demand an explanation—but any reasonable theory must account for the *survival* of so many other organisms through the same period, including mammals, flowering plants, insects, crocodiles, snakes and lizards on the land, and most kinds of mollusks,

corals, crabs and many more in the sea. A planetary cataclysm would be less discriminating—and a theory that can account only for the dinosaurs' extinction is not enough. There are probably more than 60 different suggested explanations, but many relate to the dinosaurs alone and so are irrelevant to the total picture. Many more are eccentric or unprovable. However, a number of facts about the late Cretaceous period are known, and in combination they may eventually yield a satisfactory and coherent explanation of the evidence in the fossil record.

The late Cretaceous saw considerable marine transgression—flooding of land areas by the sea—in many places. This must have reduced the space available to terrestrial animals. During the same period the supercontinent of Pangaea was actively splitting up, causing extensive changes to the earth's climate and geography. The end of the Cretaceous also included a period of drastic change in the distribution of water masses in the ocean. Finally, it has recently been shown that there was probably an enormous meteoric impact at this time (see pp34–35). A fashionable explanation attributes all the extinctions to this one event—a real catastrophe. But it is

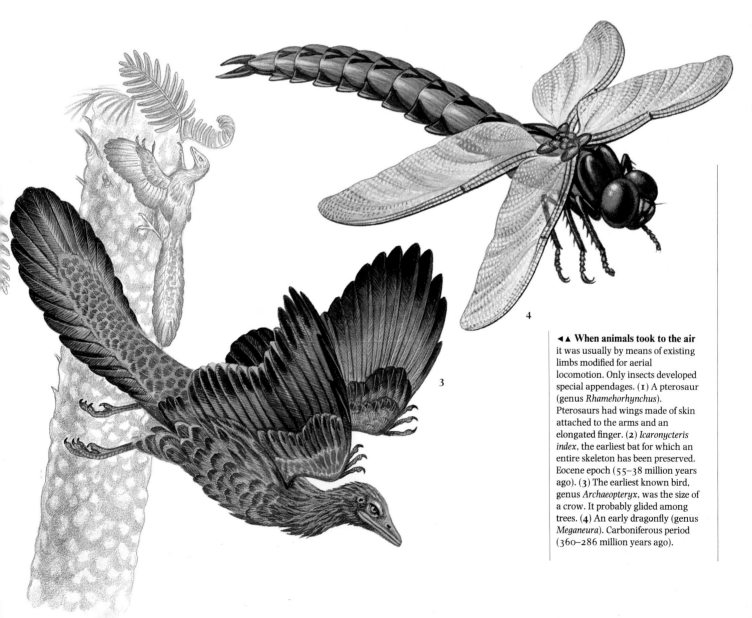

◄▲ **When animals took to the air** it was usually by means of existing limbs modified for aerial locomotion. Only insects developed special appendages. (**1**) A pterosaur (genus *Rhamehorhynchus*). Pterosaurs had wings made of skin attached to the arms and an elongated finger. (**2**) *Icaronycteris index*, the earliest bat for which an entire skeleton has been preserved. Eocene epoch (55–38 million years ago). (**3**) The earliest known bird, genus *Archaeopteryx*, was the size of a crow. It probably glided among trees. (**4**) An early dragonfly (genus *Meganeura*). Carboniferous period (360–286 million years ago).

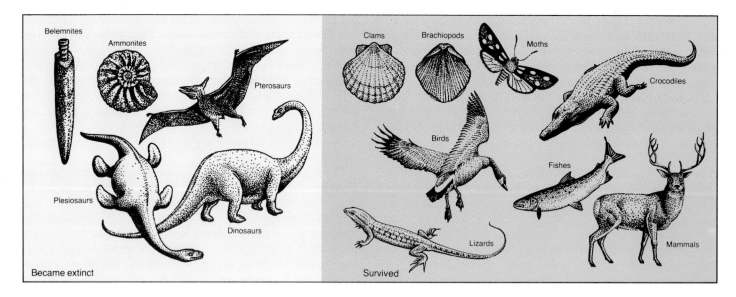

Became extinct   Survived

difficult to explain the varied survivors by this theory, and it has been found that apparently similar meteor impacts at other periods did not cause extinctions.

Recently it has been suggested that the marine fossil record shows fairly regular extinction events every few tens of millions of years. They alternate with quiet periods when nothing but a slow "turnover" of species was taking place. These extinction events were on a smaller scale than the major event at the end of the Cretaceous period; they are simply periods of increased extinction, usually followed by an increase in new species. It seems that the history of life is rather like our own: short periods of crisis alternate with times of placidity.

Growing and Changing

The story of evolution is also recorded in another way—in the life histories of individual organisms. It was once thought that the growth of the embryo actually reenacted, or recapitulated, the evolutionary history of an animal (see p88). A human embryo, for instance, seems to pass from a single-celled stage through a fishlike stage to a recognizable primate. This is incorrect.

The truth is that organisms related in an evolutionary sense look very similar to one another at an early stage of development, even when the adults are very different. All mammals reveal their common ancestry in their embryos, even with adults as different as camels and seals. Major groups of marine organisms often have similar larvae that grow up into very different-looking adults. This is not to say that the ancestor of the group would be identical to the common larva, although the ancestor would also have had a similar larval stage. The plan of development is laid out in the genetic code, and the genetic code of related animals is sufficiently similar to cause similarities in the early growth stages. Some of the most striking variations—great difference in size, for instance—can be "added in" to this genetic program at quite a late stage. Closely-related animals will have all but identical developmental histories. The distinctive features of the species will only appear at the end, or the species difference may be due to an acceleration or slowing down of some aspect of development.

The growth of an organism is seldom just a simple matter of getting bigger. Very few animals grow by scaling up the proportions they had as juveniles. Change is the rule: one part grows faster than another to produce the well-adapted adult. For instance, human beings would be very odd-looking animals if our adults were simply "grown-up" babies. Instead, of course, our limbs extend proportionally while the head (and our most vital organ, the brain) changes less strikingly. In nature, most growth is of this non-uniform or allometric kind, even among the invertebrates. Shape changes are often necessary to cope with the problems produced by an increase in volume. A tiny planktonic individual may be perfectly viable with a spherical shape—a larger adult may have to flatten or develop spines to cope with spreading its weight on the sea floor.

◄ **The greatest elimination of living forms** in history occurred at the end of the Cretaceous period (about 65 million years ago). A wide range of both terrestrial and marine animals became extinct. Others survived to inherit the earth.

▼ **The legacy of limited development.** In the fossil record some animals seem to be related to the immature forms of earlier animals. They may have originated as the offspring of immature forms that became sexually mature. This is plausible because some larval forms do become sexually active, as has happened in the axolotl (*Ambystoma mexicanum*), a species of salamander.

Lineages: Reading Evolution from the Rocks

In cases where the fossil record is particularly complete, and many specimens can be collected from well-exposed sections of the same rock type, it may be possible to "read" the story of evolution directly from the rocks. A series of fossil species can be placed in evolutionary order, and all being well the same sequence can be found in the same order elsewhere. This kind of evolutionary case history is often called a lineage. Of course, lineages are most often found among groups that have a good fossil record, especially marine invertebrates. The smaller the fossil, the easier it is to collect it in large numbers, so the most reliable lineages are described from a range of small fossil animals. These are often encountered in boreholes, so lineages have become particularly important in the oil industry, where they are used as a reliable means of dating rocks. Many of the changes found within a given lineage are slight—dramatic jumps in evolution are rarely recorded in this way. Sometimes different lineages at different times repeat rather similar changes, producing species that may look almost the same, even if they are totally unrelated.

Lineages are the nearest approach in the fossil record to evidence of evolution in action. A good example is that of changes within the single-celled planktonic foraminiferans: these tiny animals can be collected in great numbers from boreholes in many parts of the

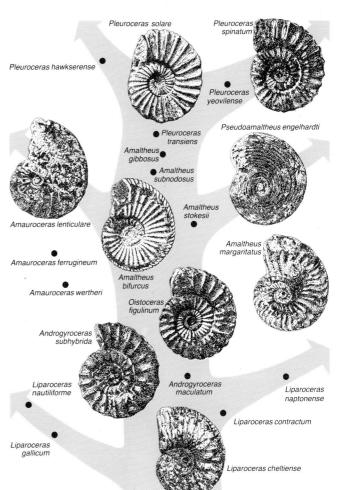

world. The same lineage is found time and time again—proof of its reality. Among ammonites, shown above, lineages are recognized on the basis of fewer specimens, although they are still numerous. Ammonites were particularly liable to go through similar lineage changes at different times, making their identification a job for the experts. Both ammonites and foraminiferans are widely used to correlate rocks from different parts of the world.

Evolution has exploited allometric growth to generate new species. Many forms seem to result from a kind of "arrested development" in which the adult of the descendant species retains one or more features typical of its ancestor's immature stage. It would be more correct to say that some of these forms show accelerated development—they become sexually mature while resembling the juveniles of the ancestral species. The axolotl, for instance, can be sexually mature while retaining external gills for breathing, an immature feature in amphibians. In appropriate circumstances, though, it will "mature" into a salamander. Some theorists have tried to explain man's unusually large brain in a similar way. Since large brain to body-weight ratios are characteristic of small mammals in general and small primates in particular, the evolution of *Homo*

can be regarded as showing the development of "immature" features. Developmental changes of this kind seem to show how profound alterations in shape can be produced with very little genetic change—a single gene might be enough to control the necessary timing. Whatever the genetic background, the fossil record shows many instances where descendant species can be related to the immature growth stages of earlier forms. In some cases these descendants are miniatures, like small growth stages that have become sexually mature. In other cases they may be normal in size, or even exceptionally large. The whole development has "spread out" to produce what are, in a sense, giant larvae. But all these distinctive species have some unique feature of their own that shows they really are separate species, and not simply the growth stage of the same species.

Evolutionary Trends

When the fossils of an animal or plant group are placed in sequence they will often show progressive changes, the result of evolutionary processes acting over a long period of time. These changes may be quite profound, affecting many different parts of the anatomy. Such evolutionary trends are not quite the same as lineages, because the different steps in the trend are not necessarily the actual record of ancestor and descendant—the real evolutionary process may have been a good deal more complicated, even including small-scale reversals of the trend. The time scale for such trends is usually large—too large for a record to be left within a single continuous sequence of rock. The trend is usually worked out by piecing the record together from different sites spanning several ages. Some trends are rather common—for instance, dinosaurs and clams both tended to increase in size. Others are specific to some particular animal or plant, usually reflecting increased specialization for a particular mode of life.

The evolution of the horse, *Equus*, from the probable ancestor of all horses, the Eocene *Hyracotherium*, involved an increase in size, but also involved a whole series of changes affecting different parts of the animal's anatomy. The toes became fewer, for instance, eventually leading to the modern hoof—effectively a single, giant toe. The brain not only became larger (which might have been expected) but acquired a more elaborate structure. The molar teeth became more complex, the neck longer, the limbs better adapted for swift running. All these changes were probably connected with the efficient exploitation of the grassland habitat to which horses are now so supremely adapted. However, the real evolutionary path of the horse is much more complex than this account of the main trends would suggest: the trends are like a simplified road map that takes no account of side roads and cul-de-sacs.

◄▼ Evolution read forwards
rather than backwards shows how
animals are constantly adapting to
new environments and to changes
in existing environments—some
with success, many ultimately with
failure. Historically there have been
numerous genera of horses or
equids: today there is only one.
Many died out because they could
not adapt to cope with new
conditions.

The modern genus (*Equus*) is the
living representative of a line that
has shown many changes,
including some long-continued
trends. The genera shown
constitute the direct line back to the
earliest known ancestor, but each of
these (except *Equus*) marked a
significant development that
resulted in other genera (see chart
BELOW).

Hyracotherium (**1**) flourished in

the Eocene in swampy forests where
it browsed on leaves. Height about
40cm (16in). *Mesohippus* (**2**), an Oligocene form,
lived in more open country where it
ran and trotted. Height about 65cm
(25in). *Merychippus* (**3**), of the
Miocene, was the first grass-eating
horse: it seems to have been
adapted to open grassland, having a
long neck with special supporting
ligaments for grazing, and leg
structures that could provide speed.
Height 100cm (39in). It was with
Pliohippus (**4**), a late Pliocene form,
that a single-toed form appeared,
the culmination of a trend for life in
open grassland. Height 120cm
(47in). *Equus* (**5**) has the faculty of
being able to cope with a range of
environments. This has enabled it to
survive elsewhere in the world after
it had become extinct in its native
America.

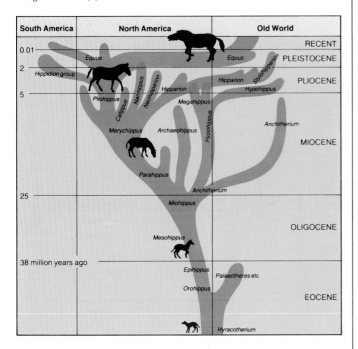

Missing Links

Much of the history of paleontology has been concerned with the search for ancestors—fossil species that connect living fauna and flora with other groups of animals or plants. Such hypothetical fossils have often been popularly described as "missing links." Because they are discovered comparatively rarely, critics of evolutionary theory often use this fact to support the argument that the idea of evolution lacks "proof." Such critics emphasize the gaps in the fossil record rather than its continuity. It is true that certain crucial fossils that would unmistakably connect one group of animals with another have proved hard to find. Perhaps this is unsurprising.

The importance of a group of animals tends to be measured by its subsequent history—the number of species or individuals within it, for instance. Yet even the greatest groups must have had humble origins, ultimately in a single species, and these early forms may well have been rare or localized. Crucial changes may have taken place in an area where there was little chance of fossil remains being preserved. These arguments suggest that most major kinds of animals and plants will tend to appear rather suddenly in the fossil record at the stage when they are starting to be successful and numerous.

However, fossil "links" are often discovered when an animal or plant group has enjoyed a period of success. This is especially true of marine invertebrate animals with a good fossil record. Such finds attract much less attention than the "missing links"—after all, it was already known that these organisms were related. These examples may seem ordinary, but that does nothing to diminish their importance. They prove that in appropriate circumstances the fossil record *can* show the likelihood that evolution has taken place. Showing the existence of real links is just as important as searching for missing ones.

The problems associated with "missing links" are well illustrated by the case of our own species, *Homo sapiens*. The newspapers regularly report discoveries of the "missing links" in our ancestry, but not all these discoveries are of comparable importance, and some are links in a chain that does not lead to man at all! In the earlier part of this century the scientific community was eagerly searching for the perfect ape-man intermediate, which led to the acceptance of a fairly amateurish hoax—the Piltdown skull—as a real fossil. Scientists are only human. They naturally want to be remembered as the discoverers of *the* missing link. But the first major advance in the search for human origins came with the shift of attention to the African continent. The best known pioneer was Louis Leakey (1903–72), who persisted in his search for many years, sustained more by his convictions than by his results. When discoveries started to be made, each in a sense was a missing link. Every fragmentary jawbone and limb illuminated a previously unknown aspect of human history. Leakey's initial discovery prompted a veritable "bone rush" into Africa, and many more sites yielded the bones of early ape-men (or man-apes). The flood of new material continues today. Again, an all-too-human enthusiasm has sometimes led to large claims for small pieces of bone, and an excess of new scientific names for the discoveries. Some of these names have disappeared in the light of more balanced assessments in the last few years.

So what of the missing link? It would be possible to regard

Soft-bodied Animals Preserved as Fossils

For the most part, the fossil record shows the hard parts of animals and plants. It is only the resistant skeletal structures, or the hardened tissues of plants, that can normally be preserved as fossils. Exceptionally, though, fossils of soft tissues *are* preserved. These rare occurrences are enormously important—they allow scientists to glimpse, however briefly, something of the true diversity of life at the time. They reveal all too clearly the inadequacy of most of the fossil record.

It is fortunate that exceptional fossils are scattered all through the geological column. In some cases the soft parts are preserved as impressions on a fine sediment, like a cast taken before the evidence could decay. In other cases the soft material is preserved as an organic film, or coated at an early stage by some substance, such as calcium phosphate or iron sulfide, which preserved a permanent record after the tissue had decayed. Sulfide preservation can be photographed with X-rays to

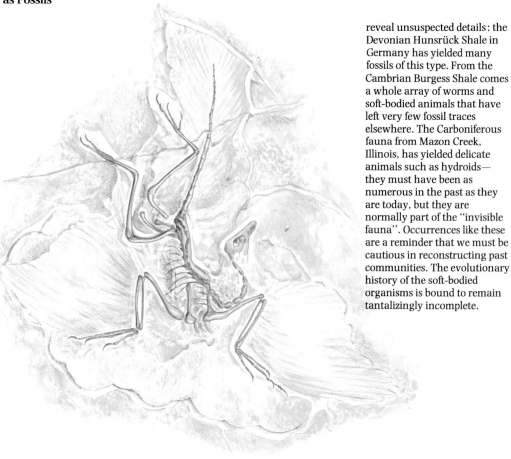

reveal unsuspected details: the Devonian Hunsrück Shale in Germany has yielded many fossils of this type. From the Cambrian Burgess Shale comes a whole array of worms and soft-bodied animals that have left very few fossil traces elsewhere. The Carboniferous fauna from Mazon Creek, Illinois, has yielded delicate animals such as hydroids— they must have been as numerous in the past as they are today, but they are normally part of the "invisible fauna". Occurrences like these are a reminder that we must be cautious in reconstructing past communities. The evolutionary history of the soft-bodied organisms is bound to remain tantalizingly incomplete.

---

the generally accepted immediate ancestor of *Homo sapiens*, *H. erectus*, as the most important link, although the somewhat older and more primitive *H. habilis* is also a candidate. On the other hand there is a vital fossil of an *Australopithecus* species that lies at or near the divergence of our own genus *Homo* from the other *Australopithecus* man-apes that failed to survive in competition with man himself. Again, perhaps the most vital link is that between our nearest *living* relative, the chimpanzee, and the whole hominid line. The more that is learned about the true complexity of evolution, the more it seems that the most important "missing link" is the one you are still searching for, the one that still lies buried in an unexplored area, or in rocks a little older than those you have examined so far. It takes a very good fossil record (and a good deal of luck) to supply the majority of the links in a chain. Even some of the best-known linking fossils, like the early bird *Archaeopteryx*, can become controversial when they are found to be less perfect intermediates than their discoverers believed. Not a link at all, say the critics, but an early "side branch" unrelated to what followed.

And so the search continues for further detailed evidence of ancestry. Perhaps the main importance of the "missing link" idea is the stimulus it gives for continued research into the problematic areas of evolutionary history. After all, a world where all the "missing links" had been discovered would be a good deal less exciting.   RAF

◄ **A link preserved** ABOVE. It was on account of the remarkable preservation properties of the Solnhofen Limestone Formation in Bavaria that a link between reptiles and birds has survived in detailed fossil form: *Archaeopteryx lithographica*.

◄▲► **A link contrived.** In 1912, near Piltdown Common in Sussex, England, a solicitor, Mr Charles Dawson, unearthed the fossilized fragments of a cranium and a jawbone. When reconstructed they turned out to be parts of the earliest human skull yet found, and, more remarkable, a link between apes and man LEFT. The find retained scientific respectability until 1954 when it was proved to be a hoax. The identity of the hoaxer, however, remains a mystery. ABOVE Dawson and Dr A. Smith Woodward, a keeper of paleontology at the British Museum, search for further remains of the Piltdown skull. RIGHT The Piltdown Men (1914–15), by John Cooke, with Dawson and Smith Woodward standing to the right of the portrait of Charles Darwin.

Living Fossils
Some survivors of ancient groups

Scattered throughout the world there are survivors from the past, the enduring relics of groups of animals or plants with an ancient lineage that may have been widespread and varied in the distant past. There are only a few of these "living fossils," but they are very important to biologists because they provide detailed information about the flora and fauna of past geological eras. Why these particular organisms should have survived is a controversial question. One popular explanation attributes it to some particular specialization—perhaps their habitat is unusually stable, or offers little real competition, allowing them to survive changes that wiped out their ancient forebears. Another, simpler, explanation is that they are lucky—the biological equivalent of the man who breaks the bank at Monte Carlo. In a world buffeted by chance events, some organisms are bound to "get lucky." A third explanation points to the fact that some of the survivors have always evolved fairly slowly. Their durability may be the result of a slow rate of change compared with other groups of animals and plants that have followed spectacular explosions in development with equally spectacular phases of extinction.

Whatever the explanation, every evolutionary biologist hopes for the discovery of new "living fossils" as an insight into the past. Even now, a more systematic exploration of the ocean depths is revealing new examples.

The famous **coelacanth** fish (*Latimeria chalumnae*), the sole survivor of its order, had close relatives in the Cretaceous period more than 100 million years ago. Many of the fossil "lobe-finned" fishes are older still, going back to the time when the first steps out of the water onto the land were being taken. Many specialists believe that the crucial transition was made by a relative of the group to which the coelacanth belongs, giving rise to all the four-limbed (tetrapod) animals. The coelacanth is not as rare as was once thought, and today its biochemistry is being investigated to test conflicting theories of tetrapod origins.

The **horseshoe crab** (genus *Limulus*) sometimes swarms in the coastal waters of North America. Fossils of an almost identical animal are known from Jurassic rocks, but the heyday of the limuloids was a good deal earlier, during the Carboniferous period some 300 million years ago. At this time there were dozens of different kinds of horseshoe crab, many living in a variety of habitats in the coal swamps. Horseshoe crabs can flip over onto their backs while swimming, and the powerful crushing bases on their legs can make short work of a clam. At one time they were thought to be close relatives of the trilobites (which have no living relatives) but any resemblance is now thought to be superficial.

The **pearly nautilus** (genus *Nautilus*) is a relative of the squid and the octopus, but it has a much longer history. Relatives of the nautilus can be found in late Cambrian rocks at least 500 million years old, making even the coelacanth seem like a newcomer. Nautiloids were very abundant and varied in the Paleozoic seas, and only a little less so during the Mesozoic. The pearly nautilus is related to the fossil ammonites, and study of the living species is one of the few ways we have of understanding how this important fossil group might have lived. The ammonites, after a spectacularly successful history, failed to survive the Cretaceous. Why the pearly nautilus alone should still be with us is something of a mystery.

The curious stumpy-legged **velvet worm** (genus *Peripatus*), found under rotting logs in moist tropical climates, is not an uncommon animal. It is, however, an extremely ancient one. Fossils of a marine animal very like *Peripatus* have often been found in Cambrian rocks at least 550 million years old. The Onychophora, the group to which *Peripatus* belongs, has probably never been very large or diverse, and may be an example of a group that has persisted almost by virtue of its own inertia. *Peripatus* offers important evidence in the debate about whether arthropods had one or several evolutionary origins. It may be related to the insects, but less closely, if at all, to crabs, shrimps, and the horseshoe crab, also shown on this page.

The beautiful **maidenhair tree** (*Gingko biloba*) is the only living representative of a group of plants that were abundant and diverse during the Mesozoic, including shrubs and trees of many different modes of growth all over the world. The last species was preserved as a cultivar in China, and has now found a new lease of life in botanical gardens worldwide.   RAF

▲ ▶ **Some examples of living fossils.** TOP LEFT a pearly nautilus (*Nautilus macromphalus*); diameter of shell about 25cm (10in). TOP RIGHT a velvet worm (genus *Peripatus*). Surviving species vary in length from 1.5cm to 15cm (0.6–6in). ABOVE a horseshoe crab (*Limulus polyphemus*). RIGHT, ABOVE a coelacanth (*Latimeria chalumnae*), found only off the Grand Comore and Anjouan islands. RIGHT leaves of the maidenhair tree (*Gingko biloba*).

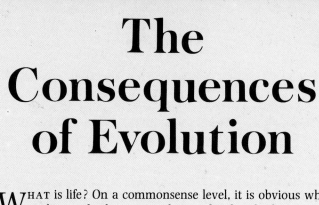

The Consequences of Evolution

WHAT is life? On a commonsense level, it is obvious what is alive and what is not; but to decide whether viruses are living (for example) is not so easy. Fortunately there are several fundamental characteristics that all living things have in common: functionally, they all reproduce and inherit; and biochemically, they are all controlled by the same system of genetic programming based on the nucleic acids DNA and RNA. This basic *homology* is strong evidence that all living things have a common ancestry.

Resemblances of form between different species do not always mean that the species are closely related. Sometimes they are merely analogies, due to similarities of function. However, there are many demonstrable cases of true homology—resemblances that are due to common ancestry. These provide useful evidence in working out the phylogenetic tree showing how species and higher groups are related to each other. Embryology—the study of embryo development—also makes a contribution here, for the early developmental stages of an embryo will generally represent ancestral embryonic forms.

As evolution takes place over the immense time-span of geological ages, the different kinds of animals and plants diverge from their ancestors. More and more new species come into existence, and the world of life becomes ever more complex and diverse—especially in the tropics where conditions are most suitable for survival and growth.

This evolutionary divergence does not happen at random. The mechanism of natural selection works in such a way that species become *adapted* to suit their way of life and environment. Successful characteristics are thus preserved and encouraged (provided they have a genetic basis). But "success" is meaningless in isolation: an animal is successful if it can survive and reproduce within its particular environment.

The adaptation of species takes two basic forms: physical adaptations, such as mimicry for avoiding predation; and behavioral adaptations such as the adoption of social behavior for mutual protection or to encourage reproduction. Sometimes adaptation is—in human terms—extraordinarily ingenious. Different kinds of cuckoos lay eggs that precisely match those of the various "host" species, for example.

◄ **In the beginning.** It has been suggested—and some experiments support the view—that life could have originated under conditions similar to those found during volcanic activity.

THE ORIGIN OF LIFE

How did life begin?... What makes a life-form living as opposed to inanimate?... Chemical experiments to simulate the origin of life point to a common origin

IN the *Origin of Species* Charles Darwin remarked that "probably all the organic beings which have ever lived on this earth have descended from some one primordial form, into which life was first breathed." Thus, although he earned a place in history for explaining how one life-form evolves into another, he evaded the important question of how life originated in the first place. Later authors, notably the Russian biochemist A.I. Oparin (1894–) and the British geneticist J. B. S. Haldane (1892–1964), have provided some suggestions as to how life might have begun, but much remains unknown. There are some who would question the idea that life originated biochemically on earth—for example the "creationists," whose extreme religious views preclude them from accepting *any* scientific explanation of the origin of life. There is also a small group of people, most notably the British astronomer Sir Fred Hoyle, who adhere to the view that life originated in another solar system and came to earth "ready-made" in the form of dormant spores.

Most biologists, however, consider that life did indeed originate in a biochemical manner, and that it did so on the primitive earth rather than elsewhere. (This view, needless to say, is not incompatible with a belief in God.) There can never be any direct evidence for this proposal, in the way that there is fossil evidence of evolution. However, it is possible to infer the conditions prevailing at the time when life is thought to have originated (about 3.5 billion years ago), to simulate those conditions in an experiment, and to see whether chemical changes occur which are similar to those that must have been involved in the origin of life.

What is Life?

Before examining the results of this kind of experiment, it is necessary to look at what is meant by "the origin of life." The phrase implies that life can be defined, and that it is possible to distinguish in some way between those entities which are life-forms and those which are not. There is no problem in making this distinction in relation to everyday objects: birds and snails are alive, while rocks and clouds are inanimate. However, other less familiar objects, such as viruses and coacervates (see below), present some difficulty. Given that such problematical objects abound in the microscopical world, it is useful to have some criteria by which to judge an entity in order to decide whether it is alive.

There are two different ways of attempting to define a living organism, and neither is without its difficulties. First, life can be defined in terms of chemical composition. For example, all living organisms contain one or more self-replicating nucleic acid molecules (DNA and/or RNA). Self-replication is a necessary, but not a sufficient, condition for a life form. That is, all organisms contain nucleic acid but not all entities containing nucleic acid are organisms. So the definition needs further refinement; but this produces other problems. For example, living organisms could be defined as those entities containing

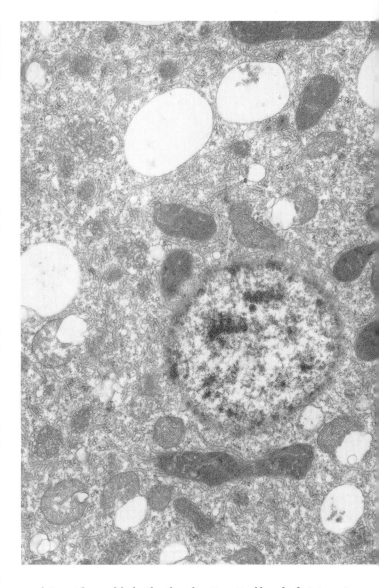

nucleic acid capable both of replicating itself and of giving rise to protein without any "outside assistance." By this definition, viruses are inanimate—a conclusion at odds with most biologists' personal intuition.

The alternative approach is to define a living organism in terms of the attributes necessary for natural selection to work. These are: reproduction, variation and inheritance. In most cases this approach will produce the same answer as the chemical one. An advantage of the evolutionary definition is that it would also be applicable to non-carbon-based life-forms (if these exist in other solar systems—a matter about which nothing is known). A disadvantage is that it is less easy to apply to the experiments described below, in which the effects are usually monitored chemically.

The best approach is probably to dispense altogether with a rigid distinction between life forms and other forms, and simply to talk of a biochemical progression: from simple organic molecules like methane (CH_4), through nucleotides and amino acids, then oligonucleotides and peptides, to loose aggregates of self-replicating nucleotides and their resultant proteins, and eventually to the membrane-enclosed equivalents of these

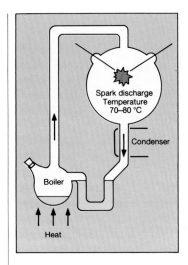

◄ **The basic organism of life** is the cell; a minute living being, able to produce energy and reproduce itself. It seems simple but in reality is a thing of extraordinary complexity, containing numerous subsidiary structures that carry out complex functions. The attainment of this level of life was a significant and crucial step in evolution.

► **The spark and the soup.** ABOVE Scientists have attempted to recreate the conditions in which life may have begun. It is thought that 4 billion years ago a "primitive soup" covered the earth. It contained such substances as hydrogen, methane and ammonia. The diagram shows an experiment conducted by Stanley L. Miller in the 1950s. The apparatus contained the substances presumed to have been in the soup. The effect of lightning was simulated by passing an electric discharge through the mixture. After a week of circulation, in which the materials were boiled and condensed, various biologically important chemicals had formed, including acids and nucleotides.

► **Recreating life,** Professor Miller with his apparatus.

▼ **Man-made cells,** "proteinoid microspheres" produced by Sidney Fox by heating organic chemicals to melting point and quenching them with cold water. They resemble what may have been the first living organisms.

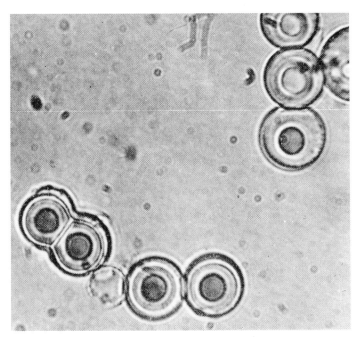

aggregates. Any process in which this whole sequence occurs can be considered as a means by which life begins, and different individuals can decide for themselves at which point in the sequence they consider the system to be alive.

The Experimental Approach

The primitive earth (see p14) is thought to have had an atmosphere consisting predominantly of hydrogen, methane, ammonia and carbon dioxide. This contrasts with the present oxygen/nitrogen-based atmosphere, the oxygen in which is largely a by-product of the photosynthetic activity of plants. Our picture of the primitive earth's atmosphere derives partly from a study of the atmospheres of neighboring planets such as Venus, where oxygen is not an abundant atmospheric gas. In attempting to reconstruct the chemical processes which occurred in the "primitive soup"—a salt solution thought to have covered large areas of the earth before the origin of life—several experimenters have set up, in the laboratory, containers with mixtures of hydrogen, methane, ammonia, water and other nonbiological substances. They have then passed an electrical discharge (simulating lightning) or ultraviolet light through the mixture, and have observed any resulting chemical changes. Such experiments have produced not only amino acids and nucleotides, but even small peptides and ATP (a building block of DNA).

Although collections of such molecules are still a long way from constituting living organisms, solutions containing peptides and nucleic acids are known to break up into droplets called coacervates in which the large molecules are concentrated. These have the remarkable property of splitting into two or more smaller coacervates, each resembling the "parental" one in chemical constitution, once they have grown past a certain size. Thus they behave in a way which resembles asexual reproduction by fission. This behavior, coupled with their variation and a degree of rudimentary "inheritance," means that these coacervates may be said to have been acted upon by natural selection. Once selection comes into play, further refinement is always possible. For example, coacervates that break up due to variation in temperature or pH might be replaced by those that do not. Ultimately such a process of selection for robustness might lead to a para-cellular arrangement whereby the internal biochemistry is protected by a membrane.

The picture that is emerging from experiments on simple molecules and from investigations of the behavior of present-day coacervates (for example, of gelatin), provides us with a reasonable simulation of *how* life orginated on the primitive earth—though no doubt future experiments will produce further refinements. One further question that arises is *how often* life originated. It seems likely from the near-universality of the genetic code—the precise formula for conversion of nucleotide triplets to amino acids—that there was only one *successful* origin of life. There is a slight complication here in that organelle genomes exhibit variations from the usual code. However, these are so slight that it is most unlikely that they represent different origins. They almost certainly arose through modification of the generalized code characterizing nuclear genomes (or vice versa) at some later, but still very early, stage of evolution.   WA

HOMOLOGY

Homology indicates common ancestry, analogy common function. . . Jaw bones in reptiles and ear bones in mammals. . . Examples of convergent and parallel evolution. . . Homology in plants. . . How can homology be distinguished from convergent evolution?. . . Homology in behavior. . . Ritual behavior in birds. . . Lorenz's ducks and geese. . . Embryology. . . Embryo development and evolutionary past. . . Early developmental and ancestral embryonic forms. . . The "left-handed" snail. . . Biochemical and physiological homologies. . . What is life?. . . The genetic program and physical form

THE concept of homology was first formally introduced into biology in 1843 by the anatomist Sir Richard Owen (1804–92) to describe the fundamental similarity of structure between corresponding organs of different animals. Following Darwin's *Origin of Species* (1859), such similarity became understood as a probable indication that the animals concerned had a common evolutionary ancestor. However, a distinction must be made between similarity due to common ancestry, or homology, and resemblance which is due solely to similarity of function, known as *analogy*. Thus the forelimbs of humans, horses, whales and birds are homologous: they are all constructed on the same pattern, and include similar bones in the same relative positions because these are all derived from the same ancestral bones. The wings of birds and insects, on the other hand, are analogous: they serve the same purpose, but do not constitute modified versions of a structure present in a common ancestor. The wings of birds and bats are homologous in skeletal structure because of descent from the forelimb of a common reptilian ancestor; but they are analogous in terms of their modification for flight—feathers in birds, skin membranes in bats.

Homologous Structures

During evolution homologous structures may come to serve very different functions, and may even bear little resemblance to each other. The bones of the vertebrate skull, especially the jaw and ear bones of reptiles and mammals, are an excellent example of this. In reptiles the hinge between the upper and lower jaws is formed by the articulation of two bones: the quadrate of the upper jaw, and the articular of the lower. The lower jaw also contains a number of other bones: the dentary, angular, supra-angular and pre-articular, among others. Both the quadrate and articular project into the tympanic cavity of the ear, but sound vibrations are transmitted from the environment to the inner ear by only a single bone, the columella auris. In mammals the arrangement is apparently entirely different. The lower jaw consists of only a single bone, the dentary, which forms the jaw hinge by articulation with a bone of the brain-case, the squamosal. However, the mammal has a chain of three bones transmitting sound to the inner ear: the malleus, incus and stapes; and the extra leverage these bones can exert on each other allows much more sensitive sound transmission than does the single auditory bone of reptiles. Fine morphological study of these jaw and ear bones in reptiles and mammals,

combined with the existence of fossils showing a somewhat intermediate condition, shows that during the course of the evolution of mammals from reptiles, the reptilian quadrate and articular, which formed the jaw hinge, have become the mammalian incus and malleus respectively, while the columella auris has become the stapes. Several other bones of the reptilian lower jaw have also changed their function; some have been lost entirely.

Homology can often be detected by morphological investigation, which sometimes reveals some very curious anomalies. The best-known example of this is provided by the course of the recurrent laryngeal nerve in mammals. (See right.)

Just as homologous structures can diverge in function during evolution, nonhomologous structures can come to resemble each other, usually as a result of adaptation to similar functions. Analogous features that have independently evolved to become more similar are said to have undergone convergent evolution. For instance, whales and fish show many convergent similarities in gross external morphology resulting from independent adaptation to an aquatic mode of life; and the eyes of squid and fish, precision instruments serving the same purpose, show very close resemblance, yet they evolved totally independently, as can be seen by studying their embryonic development.

Convergent evolution is not always distinct from parallel evolution, in which similar features (both homologous and analogous) evolve along the same pathway in two or more lineages with common ancestry. Marsupial evolution in Australia paralleled the evolution of placental mammals elsewhere. There are, or were, marsupials in Australia resembling wolves, mice, flying squirrels and moles. Some of their similarities are due to homology since they are all mammals, but others have resulted from parallel adaptation to similar modes of life.

Examples of homologous but superficially very different features are not confined to animals, nor solely to morphological structures. For instance, the spines of cacti are homologs of the leaves of other plants. However, while cacti are an entirely New World family, there are many examples of cactus-like succulent plants in the Old World (for instance some African euphorbias or spurges) which show convergent evolution of a number of different features producing a similar adaptation to life in arid environments.

In the construction of evolutionary trees or phylogenies it is the homologous features which provide information about the relatedness of different organisms. However, it is sometimes difficult to decide whether similarity is due to homology or convergence. In general, analogous similarities resulting from convergent evolution seldom show resemblances in fine detail. Also, if two organisms resemble each other in many features, it seems more likely that the similarity between any one pair of features is due to homology and thus to common ancestry. Some classes of characters may be more valuable than others as a reliable indication of common ancestry. Darwin himself considered it a general rule that the less any part of an animal is used for specialized purposes, the more important it becomes for classification. In particular he indicated the value of vestigial organs. Such organs are common throughout nature: for instance, in many snakes one lobe of the lung is vestigial, and in others there are vestiges of the pelvic girdle and hind limbs; some whales possess fetal teeth and hair which subsequently disappear. If these vestigial features have no function and are no longer subject to strong natural selection, homology between them will be unclouded by adaptation to particular purposes. RHC

Homology in Behavior

The idea of homology can also be applied to animal behavior, ie similar behavior patterns in different groups of animals can be considered as evidence for common ancestry. However, applying the idea of homology to behavior is no simple matter, for several different reasons. Perhaps the most obvious is that behavior leaves fewer fossils (see Trace Fossils, p10). It is simply not possible to see how ancestral forms behaved in the way that it is possible to examine the limbs of early vertebrates and see how these evolved into the very different structures seen today. With behavior all that can be done is to compare existing species. But this too has its difficulties. Behavior patterns may evolve so as to resemble each other because they serve the same functions, like the wings of birds and insects, rather than as a result of common ancestry. Indeed, behavior is often a very flexible feature of an animal species: it is capable of rapid change in response to changed environments, while modification through learning is possible within the lifetime of each individual animal. As a result, animals which behave in a similar way may often be those which are faced with similar problems and whose behavior has converged, rather than those which have evolved from a common ancestor. Often it is possible to see similar behavior in animals known not to be closely related, simply because they are faced with similar problems: for example, they may be meat-eaters which must capture their prey, or migrants which must navigate to their winter quarters.

Despite these difficulties, certain aspects of behavior do split up in ways that suggest homology. For example, birds can scratch their heads in two different ways, and all members of a particular species use the same technique. They may either keep the wing close to the body and raise the foot to the head outside it, or they may move the wing out and bring the foot

▲ **An arch anatomist, caricatured,** Sir Richard Owen. He formulated the concepts of homology and analogy, but fiercely attacked Darwin's theory of natural selection.

▼ **A good example** of how a basic structure has changed in form and function is the evolution of the mammalian ear. Three bones that form part of the hinge of the reptile jaw (articular, quadrate, columella auris) have evolved into a three-bone link for transmitting sound to the inner ear (consisting of malleus, incus, stapes).

▶ **A curious anomaly.** During evolution a structure is sometimes carried over into a new group or species without itself being "redesigned." An interesting example of this is found in the homology of certain nerves and blood vessels of fishes and mammals.

In fishes, which are comparatively ancient in evolutionary terms, branches of a nerve from the brain (the vagus nerve) loop around each of the 3rd, 4th, 5th and 6th blood vessels which run between the gill slits.

Only two of these branches remain in mammals, as the anterior and recurrent laryngeal nerves, connecting the brain to the larynx. However, the recurrent laryngeal nerve still loops round the remnant of the 6th arterial arch, now known as the ductus arteriosus; so from the brain to the larynx the nerve runs down the neck, round the ductus and back up the neck. This nerve is far longer than it needs be to connect the brain and larynx. In the giraffe the nerve is about 4.5m (15ft) long.

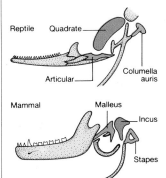

Reptile — Quadrate — Columella auris — Articular

Mammal — Malleus — Incus — Stapes

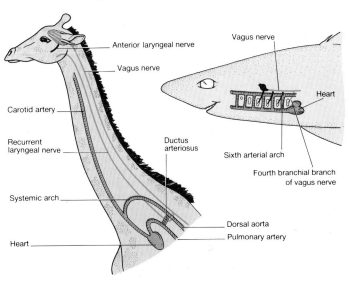

Anterior laryngeal nerve
Vagus nerve
Carotid artery
Recurrent laryngeal nerve
Systemic arch
Heart
Vagus nerve
Heart
Ductus arteriosus
Sixth arterial arch
Fourth branchial branch of vagus nerve
Dorsal aorta
Pulmonary artery

THE PENTADACTYL LIMB

▼ **The pentadactyl limb,** meaning a limb with five digits, is a characteristic feature of the tetrapods, or four-footed animals, ie amphibians, reptiles, birds and mammals. The basic form of the pentadactyl limb (in the pattern of bones (shown here), muscles and nerves) is similar in all four-footed animals, past and present, and the limbs of all tetrapods appear to be built to a similar design in both fore and hind limb. This was even true of the earliest tetrapods in the fossil record, namely the amphibians known as labyrinthodonts.

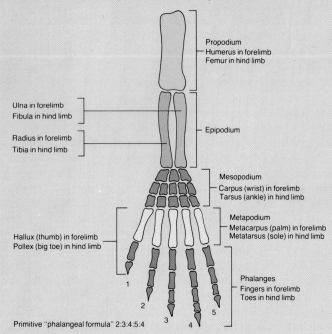

Propodium
Humerus in forelimb
Femur in hind limb

Ulna in forelimb
Fibula in hind limb

Radius in forelimb
Tibia in hind limb

Epipodium

Mesopodium
Carpus (wrist) in forelimb
Tarsus (ankle) in hind limb

Metapodium
Metacarpus (palm) in forelimb
Metatarsus (sole) in hind limb

Hallux (thumb) in forelimb
Pollex (big toe) in hind limb

Phalanges
Fingers in forelimb
Toes in hind limb

1 2 3 4 5

Primitive "phalangeal formula" 2:3:4:5:4

▼ **From fish to land animal.** The structures of fish fins and tetrapod limbs appear quite different. However, the two are comparable in basic structure. The fish ancestors of land vertebrates are considered to be crossopterygian (lobe-finned) fishes, not dissimilar from lungfishes alive today. The arrangement of the bones in the fins (**1**) comprised a single-element propodium and double-element epipodium, as in the tetrapod limb. However, various arrangements of the outer fin, known as basals and radials, were present from which may have arisen the mesopodium, metapodium and phalanges. While the inner part of the limb is similar to that of a number of different lobe-finned fishes, the form of the outer limb suggests that only one line of lobe-finned fishes was ancestral to the subsequent radiation of tetrapods. Shown also here are the bones in the pectoral girdle of (**2**) an ancient fossil amphibian and (**3**) an ancient reptile.

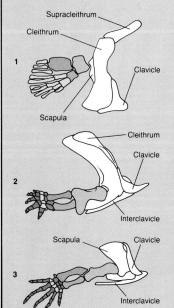

Supracleithrum
Cleithrum

1

Clavicle

Scapula

Cleithrum

Clavicle

2

Interclavicle

Scapula
Clavicle

3

Interclavicle

▶ **Running adaptations.** The hind limbs of both (**1**) mammals and (**2**) birds show loss of some digits and elongation of metatarsals (fused in birds, not in mammals) and of the tibia and fibula, although the fibula is lost in the bird. However, the axis of articulation through the tarsus differs between mammals and birds. In the mammal it is between the tibia and the tarsus. In the bird it is in the middle of the tarsus, with the distal tarsal fused to the metatarsal and the proximal tarsal fused to the tibia. This difference can be traced back to the early split of potential reptile ancestors of the mammals and birds.

Shown ABOVE are a herd of zebras (*Equus burchelli*), a typical running mammal, and ABOVE RIGHT ostriches at full gallop, a species of bird that totally relies on its legs to move around its habitat.

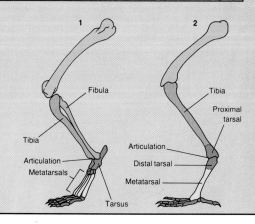

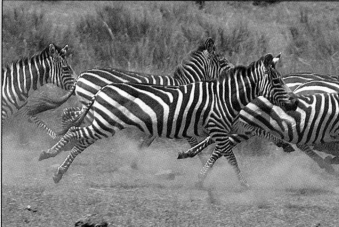

1

Fibula

Tibia

Articulation
Metatarsals

Tarsus

2

Tibia

Proximal tarsal

Articulation

Distal tarsal

Metatarsal

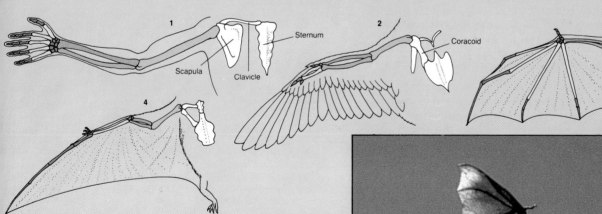

▲ Degrees of homology and analogy. A human arm (1) is *homologous* to a bird's (2) wing because, as can be seen from their development and from the fossil record, they are both formed from the same original structure. However, a bird's wing is not *analogous* with a human arm: the bird uses the wing for flying, the human uses the arm for grasping, and other mammals use the forearm for locomotion on land. The bird's wing is analogous to the wing of an insect, but it is not homologous. A bird's wing is both homologous and analogous to the wings of other flying vertebrates, such as pterodactyls (4) and bats (3), like the Indian flying fox (*Pteropus giganteus*) RIGHT. However, the homology here is of a somewhat restricted nature: the wing is formed from the humerus of the forearm in all three cases, but the exact way in which the bones are modified to form a structure capable of lifting the animal off the ground is somewhat different in the three types of fliers. The difference shows that they must have evolved wings independently from different lineages of flightless vertebrates.

Sternum

Scapula

Clavicle

Coracoid

up to the head from beneath it. Among the wading birds, sandpipers adopt the latter method and plovers the former: oystercatchers behave like plovers, and this is one piece of evidence which has been used to suggest that they are more closely related to plovers than they are to sandpipers.

Other aspects of behavior which have been examined in this way, again mostly from birds, are the displays which animals use when courting. Although these may sometimes show convergence (for example, animals living in brightly-lit places tend to use striking and colorful postures), this is much less likely than in features such as feeding techniques or responses to predators. The main function of many displays used by courting males is to attract a mate and to indicate to her clearly and unambiguously that the displayer is a fit and suitable member of her own species. Displays generally differ between species, but the differences tend to be arbitrary and to become progressively greater as species become less closely related, rather than showing any convergence. After all, if the main point is that species should not attempt to interbreed, why should unrelated species use the same display postures?

The most famous study of the displays of related species was that of Konrad Lorenz on ducks and geese. Lorenz argued that the more closely related species would tend to have more displays in common, and that displays found in only a few species would tend to be those more recently evolved and present in their common ancestor. His results corresponded well to classifications made on the basis of physical structure. Some displays, such as the piping noise made by lost chicks, are found in both ducks and geese and thus appear to be very ancient; other, more recent, postures are found only in one group or the other, or just in a small set of closely-related species. Thus displays, like structures, can be used to assess relationships between species.

As mentioned earlier, one of the problems with studying the evolution of behavior is that there is no fossil record to suggest how changes took place. However, comparison between the behavior of related species alive today can sometimes suggest probable evolutionary routes. A particularly striking example is that of empidid (dance) flies. In one nectar-feeding species,

▼► **A classic case of convergence** is that of the evolution of placental and marsupial mammals. In North and South America and especially Australia marsupial mammals (ie ones that give birth to immature young which then develop in a pouch) have occupied niches filled elsewhere by placental mammals. In so doing they have evolved similar body forms. Some representative pairs of placental and corresponding marsupial mammals are shown. (**1a**) A wolf (genus *Canis*) and (**1b**) a thylacine or Tasmanian tiger (*Thylacinus cynocephalus*). (**2a**) An ocelot (*Felis pardalis*) and (**2b**) a quoll (genus *Dasyurus*). (**3a**) A Wood mouse (*Apodemus sylvaticus*) and (**3b**) a mulgara (*Dasycerus cristicauda*). (**4a**) A flying squirrel (genus *Glaucomys*) and (**4b**) a glider (genus *Petaurus*). (**5a**) A Giant anteater (*Myrmecophaga tridactyla*) and (**5b**) a numbat (*Myrmecobius fasciatus*). (**6a**) A California or Broad-footed mole (*Scapanus latimanus*) and (**6b**) a Marsupial mole (*Notoryctes typhlops*).

2b

2a

4a

5a

5b

1a

1b

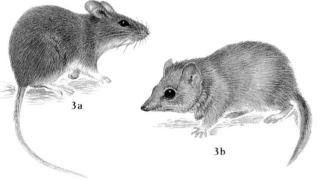

3a

3b

4b

Hilara sartor, males spin balloons out of silk and attempt to attract mates in swarms by presenting them with these apparently useless tokens. The courtship behavior of other empidids provides an explanation of how this strange ritual may have evolved. Many species are carnivorous, and in some of these the males kill flies and feed them to their females during mating. This probably enables the female to lay more eggs and so benefits them both. In other species various gradations are found between this habit and that of balloon presentation: the fly may be immobilized by wrapping it in silk; it may be sucked dry by the male before being wrapped up; and in another nectar-feeding species, the balloon is not based on a specially-killed fly but on a piece of a fly which has been picked up. This spectrum of behavior suggests that balloon presentation evolved from a type of courtship feeding which benefited the male by enhancing reproduction, but that in many species it has become no more than a token ritual which persists because males that fail to follow it cannot obtain mates.

Thus while homology in behavior is hard to study because there is no access to the past, there are instances where comparison between present-day animals can give us a plausible insight into the changes which must have taken place to generate the behavior patterns we see today.   PJBS

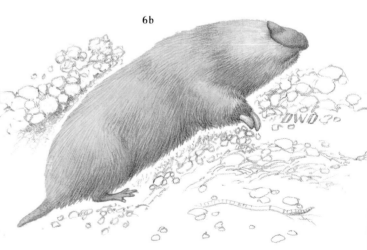

6b

6a

Embryology

It has been recognized for some time that there is a relationship between the pattern of long-term evolutionary change and the sequence of very short-term embryological changes through which an individual animal develops. The basic nature of this relationship, about which there is no real disagreement, is that two types of organism sharing a common ancestor are usually more similar during early developmental stages than they are when they have reached adulthood. For example, both primate and fish embryos possess gill slits but, while adult fish retain these, adult primates clearly do not.

While the existence of a general relationship of this kind is not disputed, the exact form of the parallel between development and evolution—or ontogeny and phylogeny—has been the subject of some debate. Two contrasting views were put forward in the 19th century by two German biologists, Ernst Heinrich Haeckel (1834–1919) and Karl Ernst von Baer (1792–1876).

Haeckel's view, which is often referred to as "recapitulation" or the "biogenetic law," is that a later-evolved animal passes through, in the course of its development, the adult stages of its ancestors. This view is now rejected on two grounds: first, recapitulation as proposed by Haeckel would mean that natural

selection could not remove or fundamentally modify the adult stage of any organism; and second, it also assumes that natural selection is able somehow to achieve a sequential "stacking" of different adult stages, which must then be condensed into the extremely short period of time that the animal is in its embryonic stage. Both of these assumptions are contrary to what is now understood about the nature of selectively-driven evolutionary change. Moreover, intensive study of the development of a particular animal simply does not reveal the adult forms of its ancestors as Haeckel suggested. For these reasons, Haeckel's theory of recapitulation has been largely rejected.

Von Baer's interpretation of the relationship between development and evolution was quite distinct from that of Haeckel—a fact which has not always been appreciated, but which has recently been highlighted by the American paleontologist Stephen Gould. Von Baer regarded the gill slits of primate embryos, and other developmental reflections of evolutionary ancestors, as representing embryonic, rather than adult, ancestral forms. Natural selection does not work by means of an evolutionary "stacking" of a series of adult forms. Rather, it modifies the adult stages of development more rapidly than the embryonic stages, which consequently bear a closer resemblance to ancestral forms. In other words, the later the

◀ **From gift to token.** A female empidid fly feeds on a bug presented by the male (*Empis livida*). In other species the present is reduced until in *Hilara sartor* the male merely presents a balloon of silk.

▶ **Embryos compared.** The embryos of four animals (tortoise, chick, dog, man) depicted by Ernst Haeckel, showing their similar features at an early stage in growth. Haeckel believed that every animal "climbs its own evolutionary tree during development," passing through the adult forms of its ancestors.

▼ **Giants of German zoology,** Karl Ernst von Baer LEFT and Ernst Haeckel RIGHT. Both men made many important contributions to zoology. Von Baer showed that mammals develop from eggs. His systematization of embryology gave the subject an important status. Haeckel suggested that inheritance was a special concern of the cell nucleus; he also spread Darwin's theory of natural selection in Germany.

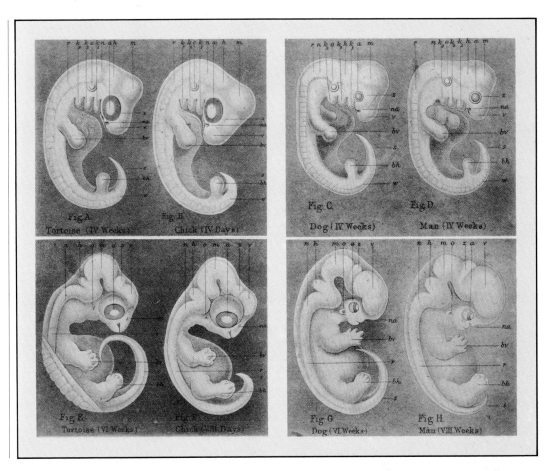

stage of development, the greater the speed of evolutionary modification.

A Modern Approach: Genes and Development

This relationship between the speed of evolutionary change and the stage of development, implicit in von Baer's approach, blends well with current views on the mechanics of natural selection; indeed, even if no comparative embryological data were available, there would be a number of theoretical reasons for assuming just such a relationship. The development of an animal is controlled by (and controls) its genes, and different developmental stages are, to a certain extent, controlled by different sets of genes. For example, in the fruit fly genus *Drosophila*

the determination of different body-segments in both larvae and adults is controlled by a gene-complex called bithorax, which comes into effect at a very early stage of the life cycle. The many other genes which take effect at a later stage control more minor developmental processes; they have no control over the main pattern of segmentation, which has already been established by bithorax, together with a few other early-acting genes.

What is the link between the genetic control of development and the slower evolution of earlier developmental stages? There are two aspects to this question. First, genes affecting early stages of development indirectly affect the later stages too, whereas the reverse is not true. Thus, if all of the genes controlling development evolve at the same rate, then it necessarily follows that the appearance, or phenotype, of later developmental stages will evolve more rapidly than that of earlier ones. However, this is unlikely to be the whole story. For the genes controlling later developmental processes almost certainly evolve more frequently than those controlling earlier ones. This is because mutations in the earlier-acting genes have a greater effect on the phenotype than mutations in later-acting genes; and, as the British geneticist Sir Ronald Fisher pointed out in 1930, the greater the effect a mutation has on an animal's phenotype, the less likely the mutation is to be selectively advantageous.

An Example of a Developmental Change in Evolution

There is, then, a three-way relationship between von Baer's interpretation of the parallel between ontogeny and phylogeny, the genetic control of development, and the operation of natural selection. However, because the commonest type of evolutionary change is the modification of late-acting genes, with less obvious effects on development, it is hard to find examples where a clear connection can be seen between the mutation of a gene, an alteration in the developmental process it controls, and a change in the adult phenotype of a sort that has actually occurred in evolution. One good example, though, is provided by the freshwater snail *Lymnaea peregra*. The shells of this species are normally dextral (that is, they form a right-handed spiral when viewed from above), but some natural populations contain a certain low frequency of sinistral (left-handed) forms. The basis of this variable phenotype is a gene which controls the orientation of cell divisions in the early embryo.

In addition to the evolution of the rare sinistral variant of *Lymnaea* through mutation of this gene, a whole sinistral family of species, the Clausiliidae, has arisen through a similar sort of genetic event. It seems likely that the divergence of the major groups of animals—the phyla and classes—involved radical mutations of other genes affecting early development. However, information on this is lost in the distant past.   WA

Biochemical and Physiological Homologies

The fundamental characteristics of all living organisms are so similar and yet differ so markedly from the characteristics of nonliving things that they provide obvious evidence for a unique origin of life on earth; ie that all the creatures alive today are descended ultimately from one common ancestor. Less than one-third of all naturally occurring elements, for example, are found in living things, and the proportions of the elements that *are* found are very different from those in nonliving things.

Although living beings are composed of a restricted range of elements, the molecules of which they are made are complex and highly varied in both structure and function. This is because carbon is almost unique among the elements in being able to form large molecular complexes consisting of chains and rings of atoms. Only silicon (to a lesser degree) shows similar properties. The carbon-based building blocks of living organisms are sugars, amino acids, fatty acids and nucleotides. These are themselves fairly complex, but they can combine in turn to form macromolecules: polysaccharides, proteins, lipids and nucleic acids.

An even more profound distinction between living and non-living things is that living creatures are organized in a highly ordered fashion. They are not merely collections of reacting organic compounds: the macro-molecules make up cells, the cells make up tissues and organs, and these in turn make up organisms. The persistence of this high degree of order was for a long time thought to be a special, even mysterious, property of life; for the rule in the nonliving world is that order and organization are unstable. According to the Second Law of Thermodynamics, entropy (disorder) should progressively increase in all irreversible processes in a closed system.

However, it is now recognized that two key features common to all biological systems allow them to remain ordered and organized without violating the Second Law of

▼ **A fundamental homology.** Many organic molecules are asymmetric and can exist in two forms that are mirror images (like human hands): right-handed or *dextro* (D) forms and left-handed or *laevo* (L) forms. In "test-tube" reactions both forms are produced but living organisms contain only L amino acids, D nucleic acids and D sugars. This strongly suggests a common ancestry for all life.

▶ **Sustaining biological systems:** a simple, schematic view of how resources supplied by photosynthesis or feeding are used in metabolism. Catabolism represents processes that break down these resources to release energy. Anabolism represents processes that build up resources to form new tissues.

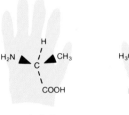

L-alanine   D-alanine

▼ **Variation in action.** There are two forms of the freshwater snail *Lymnaea peregra*. The normal form is dextral, ie the shell is coiled in a clockwise manner (center band, right). The other, sinistral, form is a mirror image of the dextral form with a shell that coils in an anticlockwise manner (lower band, right).

The form of an adult snail can be determined as early as the first division of the fertilized egg and is very obvious at the end of the third wave of cell division (ie the eight-cell stage, shown here). Embryos in which the quartet of small cells is shifted clockwise in relation to their corresponding large cells (center band) give rise to dextral adults; those with an anticlockwise shift give rise to sinistral adults (lower band). Which developmental pattern occurs is determined not by the individual's own genes but by those of its mother.

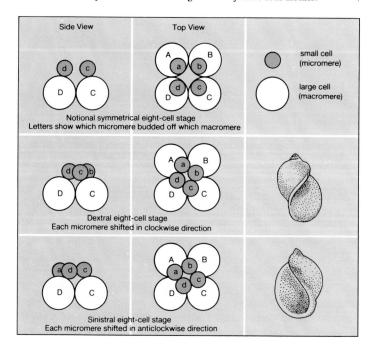

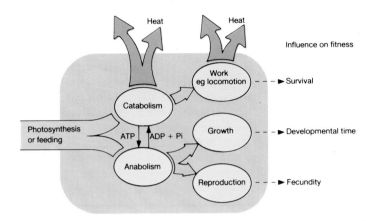

Thermodynamics: first, the systems are genetically *programmed*, and second, the biological subsystems these programs control are *open* to input and output of matter and energy.

At the most fundamental level the genetic program controls the properties of proteins by specifying the types and sequences of different amino acids from which the proteins are built up. Of the vast number of amino acids that have been synthesized only about 20 occur in living organisms. However, with a simple linear chain of only 100 amino-acid molecules chosen from these 20, the number of possible different configurations is 20^{100}—an enormous number! In practice most proteins consist of substantially more than 100 amino-acid molecules. This explains the great diversity of proteins. They include the all-important enzymes that control all metabolic processes, and other molecules that make up the fabric of the living cells. All are precisely determined by the genetic program.

Like a computer program, the genetic program has to be embodied in a physical form—it is coded as sequences of nucleotides in nucleic acids, usually DNA (deoxyribonucleic acid). In DNA there are only four different kinds of nucleotide (adenine, guanine, cytosine and thymine) and so one nucleotide cannot specify the 20 amino acids. A pair of nucleotides is also insufficient, as there are only 16 possible combinations (4^2). But there are 64 (4^3) different combinations of three nucleotides—more than enough to specify all the amino acids. This triplet code has been found to be universal. What of the 44 extra combinations? Some are duplicates, so that more than one triplet of nucleotides may represent the same amino acid; other combinations act as "punctuation."

The DNA molecule consists of two complementary strands of nucleotides held together by chemical bonds and wound into the famous double helix. In this form the information is protected against corruption. However, when the information is used to make a new protein molecule, part of one of the DNA strands is unwound and laid bare as a template that determines the sequence of amino acids in the protein.

The so-called "central dogma" of molecular biology states that information flows from genotype to phenotype, and not the other way round. In other words the genetic program specifies the basic form of organisms and how they function, but changes to the form and functioning do not influence the genetic program, and hence are not inherited. This nonreciprocity effectively rules out Lamarckian evolution (the inheritance of acquired characteristics); the genetic information is protec-

ted from changes occurring in the phenotype, most of which are due to damage and are likely to impair survival. Occasionally, of course, the genetic material mutates and this may be subjected to natural selection on the basis of the change it effects in the phenotype.

Within organisms, though, biological order and organization persist, despite the general tendency towards greater disorder, because proteins are assembled according to a program of instructions which is relatively well protected from change. The proteins formed in this way are used as building blocks to form new body tissues, to replace damaged tissues, and to form gametes (sex cells) for reproduction—the ultimate guarantee of the long-term survival of genetic programs and the organisms they specify. These processes use raw materials imported from the surroundings. It follows that there has to be an input of matter and energy (light energy in plants and chemical energy in animals) and an output of entropy, usually manifested as low-grade heat and excreta. The resource input is used not only to provide building blocks, as already described, but also as fuel to power all the active processes of the body. Phosphorylated nucleotide molecules, particularly ATP (adenosine triphosphate), act as universal stores and vehicles of chemical energy in these processes.

These universal aspects of physiology are summarized in the accompanying diagram. The resource input to organisms has to be limited by the processes and structures involved in feeding and photosynthesis. Hence resources are always limited and have to be allocated amongst the conflicting physiological demands. This allocation is controlled by gene-determined enzymes and in turn influences survivorship, fecundity and developmental rate—all of which are key components of neo-Darwinian fitness. Hence though the physiologies of all organisms are based on a common organization, they will have been "fine-tuned" by natural selection according to the ecological circumstances in which they operate.

It is interesting to ask why all biosystems have the same basic structure. The persistence of a dynamic and ordered system is probably only possible on the basis of the kind of programmed open system described above: the program has to be protected, the phenotype open, and there has to be nonreciprocity between the two. But the details of this general system—its carbon basis, the optical properties of its constituents, the specific form of the genetic code, the limited variety of amino acids, the use of ATP as a universal energy store—could be due to the chance association of these features with the one, original biosystem that acted as an ancestor for all others. Given the fundamental organization of this biosystem—programmed replication from limited resources obtained from the outside world—there had to be evolution by natural selection, those programs spreading which could transform the limited resources most successfully into replicas of themselves. However, tantalizing questions involve the extent to which particular traits are due to the accidents of history or the necessities of selection, and the extent to which the features that have arisen by accident *constrain* the possible effects of natural selection. These issues are currently taxing evolutionary biologists, particularly those interested in the molecular and physiological features of organisms.

PC

ADAPTATION

What is adaptation?. . . Convergent evolution and parasitism as examples of adaptation. . . Fleas and cuckoos: adaptive parasites. . . Behavioral adaptations in Black-headed gulls. . . Social behavior and predators. . . Great tits' territories and Wood pigeons' flocks. . . Mimicry. . . Ecological adaptations. . . Competition and ecological niches. . . Animal and plant food webs. . . The diversity of life in the tropics. . . Central and North American bats compared

NATURAL selection occurs because individuals vary in their ability to survive and reproduce. The characteristics that show these variations often have a hereditary basis; when they do, successful individuals can transmit the features that have ensured their success to their offspring. Natural selection therefore causes the evolution of successful characters, and over evolutionary time species' characteristics become molded to suit their way of life and their environment. Such useful or valuable characteristics are called adaptations.

Many lines of evidence point to the evolution of adaptations. One example is evolutionary convergence, where quite unrelated organisms following similar ways of life evolve very similar characteristics: whales and some fish are powerful swimmers, and both groups have evolved a streamlined body with attached hydrofoils, suited to this means of propulsion. The shared characteristics caused by evolutionary convergence are said to be analogous (see p82). The other major source of evolutionary similarity is relatedness, and in this case the similar structures are said to be homologous.

Natural selection does not produce perfect adaptations; it simply does the best it can with the material available. Nonetheless, some creatures show truly remarkable and often bizarre adaptations to their particular life-style. This is especially evident among parasites.

Parasitism is hard to define, but can be broadly described as a close association between two organisms to the detriment of one of them, known as the host. Some parasites, like malaria and tapeworms, live inside their hosts (endoparasites); others, like lice and fleas, live on the outside (ectoparasites).

Adult fleas usually live on birds or mammals, sucking blood from the skin with their pointed mouth parts. Unlike most insects they are wingless, probably because wings would impede progress through fur or feathers. They have immensely powerful legs, enabling them to jump onto the host from the nest where the juvenile fleas are found. Perhaps the most remarkable flea adaptation is their method of ensuring transmission between host generations: they synchronize their reproduction with that of the host by using the host's own reproductive hormones. Rabbit fleas, for instance, usually live on the ears of the rabbit. When a female rabbit becomes pregnant the fleas detect the hormones in her blood, and tend to accumulate on her. About ten days before she gives birth she produces another hormone which is detected by the fleas and triggers egg maturation in the female fleas. The eggs mature the day the young rabbits are born, and the fleas pass onto the newborn young, feed voraciously, mate and lay eggs.

After about 12 days the fleas return to the mother, and if she then becomes pregnant again they in turn start another breeding cycle.

Cuckoos are brood parasites: they lay their eggs in other birds' nests, and the young are reared by the host species. The European cuckoo parasitizes several different host species, and perhaps its most remarkable adaptation is the exact match between its eggs and those of its host. This is particularly impressive because each host species lays differently colored and patterned eggs. How does the cuckoo match its eggs to each host species? It turns out that there are different genetic forms of the cuckoo, each laying a different colored egg matching those of one host species. The female cuckoos know which host species to parasitize because they learn the habits and appearance of the host during their own early lives, when they are reared by the host parents. As adults, the cuckoos look for the same host species when laying their eggs. So the perfect match to the eggs of each host species is generated by a combination of genetic inheritance and learned preference for the host. Presumably in the past, as now, the hosts threw out any poorly matching cuckoo eggs, and this is how the adaptation must have evolved.

Behavioral Adaptations

The way an animal behaves is as important to its survival as its color patterns and its anatomy. Natural selection has molded behavior to adapt it to the life-style and environment of the species.

Many animals have evolved cryptic coloration as an adaptation against predation. For instance, many ground-nesting birds are amazingly well camouflaged, as are their nests and eggs. However, when the young birds hatch, the empty eggshell with its pale lining is extremely conspicuous in the dark nest. The shell could easily signal the presence of unhatched eggs or chicks to predators, which may be the reason why the parents of many species remove the eggshells to some distance from the nest. The adaptiveness of this behavior in reducing predation has been shown in an elegant but very simple experiment with Black-headed gulls.

Black-headed gulls nest on the ground in dense breeding colonies. The adults are a rather conspicuous black and white, but the eggs are beautifully camouflaged, and when they hatch the shells are quickly removed by the parent. The experimenters made a group of artificial Black-headed gulls' nests, each containing a hen's egg painted to look like the camouflaged gull's egg. A broken eggshell was placed 5cm (2in) away from half of these nests, while the other half had no eggshell. The experimenters left the nests to the attentions of the local foxes, crows and hedgehogs for a few days, then returned to record the results. They found that the hens' eggs had disappeared from most of the nests with an eggshell nearby, whereas in the nests with no eggshell the majority had survived. The

► **Cuckoo in the nest.** An adult willow warbler (*Phylloscopus trochilis*) feeds a fledged cuckoo (*Cuculus canorus*) which it has reared as if it was its own offspring. Cuckoos are examples of the bizarre adaptations that natural selection can produce.

experiment neatly demonstrated that a nearby eggshell increases the risk of a predator finding the nest, so the behavior of the parent gulls in removing the eggshell is highly adaptive.

The pattern of social behavior found in different animals is also a very important aspect of adaptation. Social behavior varies enormously. Some species are colonial while others are solitary. Some are highly territorial, while others can wander into close proximity without any sign of trouble.

Great tits are small, insect-eating birds found in woodlands and gardens. For most of the year pairs live in vigorously defended territories of about 1ha (2.5 acres). The pair spend all their time in the territory, and one important function of this territoriality is undoubtedly the defense of an adequate food supply. Another seems to be the reduction of nest predation. Predators rapidly learn the characteristics of their prey. After catching a prey item they form a "searching image" of the prey, a sort of mental picture of it, so that any similar prey item in the area is at serious risk from the predator. It pays similar-

looking prey items to be as far away from one another as possible, and this includes Great-tit nests. Weasels are very important predators: they enter the tree-hole nests of the Great tits and steal eggs or chicks. Studies of weasel predation on nests have shown that a nest is at far higher risk if there is another nest nearby than if the nearest nest is far away. The Great tit's territorial behavior may be partly an adaptation that ensures spacing between nests to reduce the risk of predation.

In a different bird species, the Wood pigeon, the risk of predation is lowered by living in large flocks. When one bird spots a predator such as a hawk flying towards the flock, it gives an alarm call, and every bird in the flock responds by flying away. It seems reasonable that if there are many pigeons in a flock, the chance of one of them spotting an approaching predator will be higher than if there are only a few birds present, so that birds in a large flock will be warned of the danger from a predator more quickly. Confirmation of this possibility has come from experiments in which a hand-tame goshawk was flown at pigeon flocks of different sizes. When the alarm call was given the goshawk was always further away from large flocks than from small ones, enabling the larger flocks to fly away with a much greater margin of safety. Thus Wood pigeon flocking is probably, in part at least, an adaptation against predators.

Mimicry

In general, predators are adept at distinguishing between prey that are safe and nutritious to eat and those that are noxious in some way. Some predators seem to have this ability built-in, without any prior experience of dangerous prey. For instance, young of the Great kiskadee, a predatory bird species, avoid the bright bands or rings of the very poisonous coral snake even if they have never seen it before. In other cases young predators learn to avoid dangerous prey from experience. This behavior has had profound evolutionary consequences for the color patterns of prey species.

▲ **Adaptations to life in the desert.** A pair of Dromedary camels (*Camelus dromedarius*) feed from sparse desert vegetation: these large animals of the desert have developed great tolerances for dehydration (under some conditions camels can lose water totalling over one third of their body mass) and the ability to survive wide daily changes in body temperature, (sometimes more than 10°C (18°F) between mid-afternoon highs and late night lows). Their temperature may rise considerably during hot daytime periods. Since they are large animals, weighing hundreds of kilograms, these temperature changes result in large reductions of evaporative water losses, to levels far below what would occur if they regulated their body temperatures as precisely as most mammals of their size. Camels have several other, less dramatic, adaptations for water conservation, including their feeding and drinking habits, their kidney function, and their respiratory systems.

▶ **Safety in numbers** ABOVE—a flock of wrybills (*Anarhynchus frontalis*) rise into the air from a New Zealand shore. The advantage of flocking probably varies from species to species. However, anti-predator behavior undoubtedly is important. Many eyes rather than two give greater awareness of an approaching predator, and each individual has to spend less time being vigilant. Also the confusion resulting from a mass escape is likely to hinder an attacker. Flocking may also have some

benefits during feeding. Large numbers of insectivorous birds may flush proportionately more prey than an individual. For plant and fruit eaters there are more individuals to search for good patches of food.

▶ **Avoiding being seen**—the same strategy in plants and animals. When life is near the limit of survival and population levels are low it is particularly important for both animals and plants to avoid being eaten if the species as a whole is to survive. Camouflage, as is seen in this desert grasshopper *Batrachornis perloides*, helps hide it from predators; similarly the succulent leaves of plants called living stones (*Lithops fulleri* from the Namib Desert, southern Africa) closely resemble the sandy soil and hardly emerge above the soil; this helps prevent the attentions of grazing animals.

In general, noxious prey are brightly colored and safe prey have cryptic coloration. The bright coloration of dangerous species has probably evolved as a warning to potential predators: it is easy for the predator to recognize the color pattern and avoid it. Cryptic coloration has the opposite effect: it makes it difficult for the predator to recognize the prey.

Interestingly though, some safe prey species are brightly colored. For instance many edible butterflies such as the African swallowtail are highly conspicuous. This is just one example of the fascinating phenomenon of mimicry, where a safe species copies, often in fantastic detail, the appearance of a dangerous species, and so fools its enemies into treating it as dangerous. This is called Batesian mimicry after its discoverer H. W. Bates, a 19th-century English explorer of the Amazon.

Laboratory experiments have produced hard evidence that mimicry really does protect the mimic from attack. One such investigation used the poisonous monarch butterfly species and its harmless mimic, the viceroy. Young Scrub jays were hand-reared without any experience of butterflies and then assigned to two groups. Birds in the first group were each given a series of viceroys, which they ate without any sign of fear or ill effects. This proved that the viceroy is indeed harmless, and that its color pattern is not inherently frightening to the birds. The second group of jays were given monarch butterflies. At first the birds attacked, but soon spat out the butterflies and rapidly learned to avoid them. When these jays were given viceroys they would not eat them. The birds' experience with the poisonous monarch protected the edible viceroy from attack.

Mimicry can obviously protect harmless prey species from attack, but it is quite hard to imagine how the adaptation can evolve. If the edible species initially looks very different from the potential model, it is most unlikely that a single new variant looking identical to the model will crop up. But is gradual evolution possible? Are sharp-eyed predators at all deceived by a poor imitation?

It seems that even a slight resemblance to a model can deter predators from attacking a mimic, as long as the model is really unpleasant. Further experiments with the Scrub jays proved this point. If jays that have learned to avoid monarchs are then given a harmless butterfly species of the genus *Limenitis* that shows only a moderate resemblance to the monarch, the jays are deterred from attacking it. The birds can mistake even a poor mimic for a model, so in principle incipient mimicry could be built up and perfected over many generations.

During his travels up the Amazon Bates also noticed that quite unrelated *inedible* butterflies often showed a close resemblance to each other. The German zoologist Fritz Müller (1821–97) provided a convincing explanation for this phenomenon, which is called Müllerian mimicry in his honor. Müller assumed that when young predators learn to avoid noxious prey species they will kill some in the process. This means that if there are several prey species, each with a different color pattern, the young predator will kill a certain number of each species in the process of learning to avoid each pattern. However, if the prey species all look the same, the predator need learn only one pattern. During this process it will kill some prey, but

5

4

◄▲ **Refinement in mimicry.** If, to deter predators, an ordinary butterfly species becomes adapted so that it resembles a poisonous or noxious species (known as Batesian mimicry), populations of the mimic species have to remain smaller than those of the other; otherwise predators will consider it to be the normal kind and protection may be lost. In Africa the female of the Mocker swallowtail butterfly has avoided this constraint on population size by developing three forms, each mimicking a different noxious species of danaid butterfly. (Males, however, tend to retain a single pattern. Variation would probably reduce the chances of some males mating successfully.) Shown here are a male Mocker swallowtail (*Papilio dardanus*, **2**), two of the model females (**1**, *Danaus chrysippus*; **4**, *Amauris niavius*) and their female swallowtail mimics (**3**, **5**).

now this number will be distributed between all the species. It therefore pays poisonous prey species to look the same. Müllerian mimicry is common among poisonous butterfly species in the tropics of Africa and South America.   LPa

Ecological Adaptations
All animals and plants require certain things from their environment. All have limits of tolerance as regards temperature and humidity, and plants vary in their ability to function efficiently under different intensities of light. All need certain elements such as carbon, nitrogen and phosphorus to build and regenerate their tissues. Plants may obtain these from the atmosphere and the soil, animals from their food. All need energy. For animals this involves the consumption of energy-rich food materials, but for most plants it can be obtained directly from sunlight by photosynthesis.

Since there are so many different kinds of plants and animals on the earth, all demanding the same basic resources, organisms often compete with one another in the course of their evolution. In a world of limited resources, the outcome of these competitive pressures has been the evolution and adaptation of organisms until individual species takes on certain "roles" in the community. Each gains access to the resources it needs in its own specialized way. This role of the organism is called its niche, and the development of varied niches by plants and animals has allowed many different types of organism to exist in the same environments.

Many things determine the niche of an animal. There is the nature of its food, for instance. Some herbivores will eat only one or two species of plant. The Chalkhill blue butterfly has a larva which eats one plant, the Horseshoe vetch, almost exclusively. Other animals may have wider food tolerances, but show very distinct preferences. The European rabbit, when feeding on heathlands, prefers gorse and heather to the Cross-leaved heath, but is prepared to eat it when in need. Animals may share a food resource, but consume it in different spatial habitats. The kestrel and the sparrowhawk will both take small birds, but the kestrel hunts in the open, while the sparrowhawk prefers to hunt in wooded areas or scrub, where it has a better chance of surprising its prey.

So a food resource can be divided between two or more species that prefer different habitats. It can also be divided between species that prefer to be active at different times of day. For instance, both kestrels and owls will hunt small mammals in open country, but kestrels hunt by day and owls by night. This does not completely eliminate competition between them: if small mammals are scarce, then hunting at different times does not solve the supply problem. However, it does reduce the amount of competition. Dividing resources in this way allows species with similar niches to share the same habitat.

Because animals have such specialized food requirements, there is a clear organizational structure among the animal species within an ecosystem. Herbivores such as caterpillars feed directly on plants and are eaten by carnivorous organisms such as ground beetles. These in their turn may be consumed by insectivorous mammals such as shrews, which form a food

resource for predators like the kestrel. But these feeding interactions are seldom simple. The shrew may feed directly on the caterpillar, or the kestrel on the beetle. The ability of an animal to switch its prey requirements according to taste or availability means that simple, linear food chains are rare in nature. More commonly there are complex, interlocking food webs.

Competition between animals is greatest when their potential niches overlap to a considerable extent. If two animals try to exploit the same food resources in the same habitat in the same way and at the same time, competition is intense. The outcome of such competition depends on the efficiency of each animal under the prevailing physical conditions of the environment. For instance, the two barnacle species *Semibalanus balanoides* and *Cthamalus stellatus* occupy the intertidal zone of rocky shores. Both feed on planktonic organisms by filtering sea water. The essence of their competition is the demand for space, because both settle from a floating larval form and remain sedentary throughout their adult lives. *Cthamalus* larvae tend to settle at a higher level on the shore, while *Semibalanus* is fairly evenly spread. The overlap is considerable, and *Semibalanus* is usually victorious in the lower regions, but above the mean high water mark of neap tides it is *Cthamalus* which survives at the expense of *Semibalanus*. The critical factor is desiccation: *Semibalanus* is much more sensitive than *Cthamalus*. So the response of the organism to the prevailing physical conditions of the environment can decide whether or not it gains access to food resources ahead of its rivals.

Not all interactions between animals are competitive: many other types of relationship exist. Perhaps the most remarkable of these are symbiotic interactions, in which both participants benefit. The caterpillars of the Large blue butterfly, for instance, have evolved a complex relationship with ants. When the ant finds a wandering caterpillar it feeds on a sweet secretion from the back of the caterpillar's body. This stimulates it to carry the caterpillar to the ants' nest. There the caterpillar feeds on the young ant grubs until it pupates. Such behavioral adaptation may have advantages for the Large blue butterfly, but there are dangers, too: it might not, for instance, be found by an ant, so the strategy carries certain inherent risks.

Flowering plants and insects have long had a mutually beneficial relationship. Many flowering plants rely on insects to transfer pollen from the anther of one plant to the stigma of another. In return the plant has to pay the "energy cost" of feeding the insect either on a proportion of the pollen itself or with a specially produced energy-rich fluid, nectar. There are strong selective pressures on the plant to produce just the right amount of extra material to satisfy the pollinator's needs. Too little and the pollinator may neglect the flower for a more rewarding one; too much and the plant may lack the energy it needs for vegetative growth, seed production, nutrient absorption, and all the other activities necessary for successful reproduction. However, some flowers have adopted bizarre techniques to avoid paying their energy dues. Some, for instance, have become structurally modified to resemble an

◄ **Fitting the physical environment.** Physical features often mold the niches of animals. On this rocky shore in the Galapagos Islands there are two zones of barnacles, black below, pale above, each species tolerating a different exposure to air. Superimposed on this zonation, large animals keep to different levels on the rock face, in this case red rock crabs, Marine iguanas (*Amblyrhynchus cristatus*) and Blue-footed boobies (*Sula nebouxii*).

▼ ► **The coevolution of flowers and insects.** The earliest flowers were probably pollinated by unspecialized casual insect visitors; BELOW a Scarab beetle (*Polystigma punctatum*) feeding on the nectar of a blossom. Later there evolved more specialized types of flowers, whose nectar could only be tapped by particular groups of insects; RIGHT a Honey bee (*Apis mellifera*) with pollen baskets on its legs visiting a quince flower. At the pinnacle of development a flower species can only be pollinated by a single animal species; INSET the Bumblebee orchid (*Ophrys bombyliflora*); this flower mimics the female of just one species of bumblebee; males attempt to mate with the flower and bring about pollination.

insect, and attract other flying insects, which land and attempt to copulate with them.

The combined development of structural and behavioral modifications by two organisms in this way is called coevolution. But there is a sense in which most plants and animals have coevolved. All fellow organisms form part of the environmental setting in which they have evolved. The possible interactions are many and varied. but each has played its part in shaping the evolutionary history of the organism.

The Varied Diversity of Life

For a very long time biogeographers and ecologists have been intrigued to know why so many more species of plants and animals live in the tropical regions than in higher latitudes. The tropical rain forests of the Amazon basin, for example, contain about two-thirds of the world's total of some 4.5 million plant and animal species. Why should this habitat be so much richer in species than any other? Clearly it has something to do with climate: moving from the equator toward the poles there is a marked "gradient of diversity," with fewer and fewer species present.

This latitudinal gradient of diversity is illustrated by the number of breeding bird species in different parts of Central and North America. Starting in tropical Panama, there are 667 species; in Costa Rica, there are 603; in Guatemala, 472; in California, 286; in Washington State, 235; and in Alaska 222. These figures are not strictly comparable because the land areas

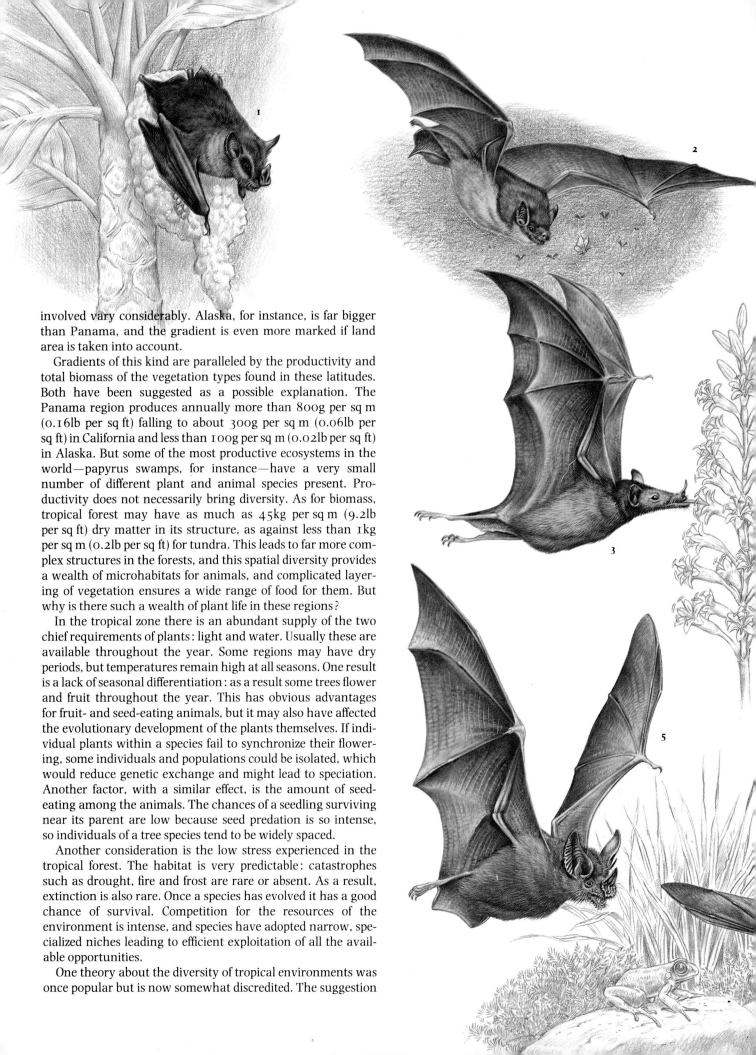

involved vary considerably. Alaska, for instance, is far bigger than Panama, and the gradient is even more marked if land area is taken into account.

Gradients of this kind are paralleled by the productivity and total biomass of the vegetation types found in these latitudes. Both have been suggested as a possible explanation. The Panama region produces annually more than 800g per sq m (0.16lb per sq ft) falling to about 300g per sq m (0.06lb per sq ft) in California and less than 100g per sq m (0.02lb per sq ft) in Alaska. But some of the most productive ecosystems in the world—papyrus swamps, for instance—have a very small number of different plant and animal species present. Productivity does not necessarily bring diversity. As for biomass, tropical forest may have as much as 45kg per sq m (9.2lb per sq ft) dry matter in its structure, as against less than 1kg per sq m (0.2lb per sq ft) for tundra. This leads to far more complex structures in the forests, and this spatial diversity provides a wealth of microhabitats for animals, and complicated layering of vegetation ensures a wide range of food for them. But why is there such a wealth of plant life in these regions?

In the tropical zone there is an abundant supply of the two chief requirements of plants: light and water. Usually these are available throughout the year. Some regions may have dry periods, but temperatures remain high at all seasons. One result is a lack of seasonal differentiation: as a result some trees flower and fruit throughout the year. This has obvious advantages for fruit- and seed-eating animals, but it may also have affected the evolutionary development of the plants themselves. If individual plants within a species fail to synchronize their flowering, some individuals and populations could be isolated, which would reduce genetic exchange and might lead to speciation. Another factor, with a similar effect, is the amount of seed-eating among the animals. The chances of a seedling surviving near its parent are low because seed predation is so intense, so individuals of a tree species tend to be widely spaced.

Another consideration is the low stress experienced in the tropical forest. The habitat is very predictable: catastrophes such as drought, fire and frost are rare or absent. As a result, extinction is also rare. Once a species has evolved it has a good chance of survival. Competition for the resources of the environment is intense, and species have adopted narrow, specialized niches leading to efficient exploitation of all the available opportunities.

One theory about the diversity of tropical environments was once popular but is now somewhat discredited. The suggestion was that tropical rain forests were not affected by the climatic fluctuations of the Pleistocene, which resulted in glaciation at higher latitudes. A number of recent studies of lake sediments in tropical sites have disproved this theory. The sediments contain pollen from past vegetation, often dating back several hundred thousand years, and show that many areas now covered with tropical rain forest experienced more arid conditions during the glacial episodes of the temperate regions.

The explanation of tropical diversity must be very complex. It involves the rich resources of the environment, its innate stability and the lack of seasonal variation. Put together, these factors have resulted in a high rate of speciation and a low rate of extinction.

PDM

◄▼ **Opportunities for Bats.** The forests of Panama in Central America contain many more mammal species than either the deciduous forests of southern Michigan or the boreal forests of the Fairbanks region of Alaska. Panama forests contain about 70 mammal species, the Michigan forests about 35 species, and the Alaskan forests about 15 species. Examining the kinds of mammals involved and the taxonomic groups to which they belong yields a surprising result: bats alone are largely responsible for the variations in diversity. In Alaska only one bat species is present, in Michigan there are seven, and in Panama 31. In percentage terms, bats comprise 7 percent of the mammal species in Alaska, 20 percent in Michigan, and 44 percent in Panama.

Just why should bats be so much more successful in the tropics than in the Arctic? The answer lies in their diets. Most bats (with some notable exceptions such as vampires) are either insectivorous (2, 6) or fruit-eating (1). However, some prey on larger animals such as frogs (5), and some are relatively omnivorous. The Alaskan and Michigan bats are all insectivores,

exploiting the abundant supplies of insects available in the temperate and arctic regions during the summer months. The Panama bats show a far greater range of food preferences: of the 31 species, 14 are insectivorous, 11 are fruit-eating, four are carnivorous and two are omnivorous.

The variety of feeding habits of bats in the Central American forests illustrates the greater number of niches, or feeding opportunities, available in the tropical environment. Fruit-bearing trees are common, and depend on fruit-eating animals such as bats to disperse their seeds. The lack of seasonality also means that food supplies are reliable and available all through the year: an animal can become totally dependent on a food resource such as fruit without risking extinction.

(1) A Mexican fruit bat (*Artibeus jamaicensis*). (2) A Big brown bat (*Eptesicus fuscus*). (3) A Mexican long-nosed bat (*Leptonycteris nivalis*), a nectar-eater. (4) A False vampire bat (*Vampyrum spectrum*), a carnivore. (5) A Fringe-lipped bat (*Trachops cirrhosus*). (6) A Mustache bat (*Pteronotus parnelli*). (7) A Fishing bulldog bat (*Noctilio leporinus*).

GRAHAM ALLEN

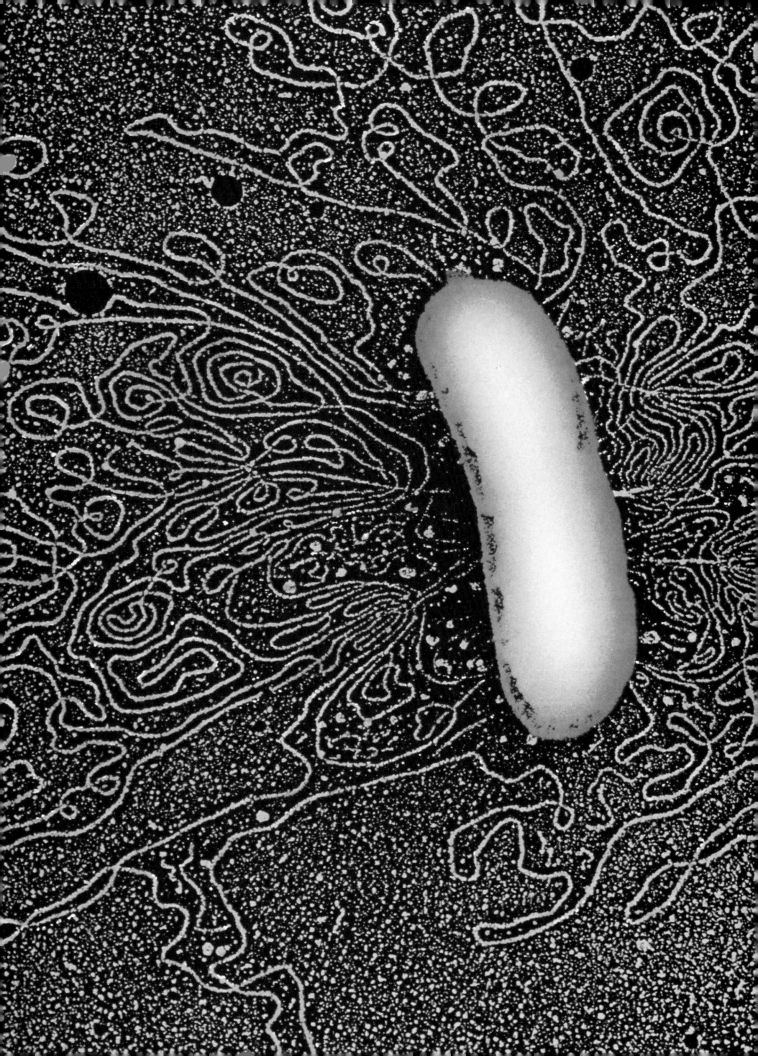

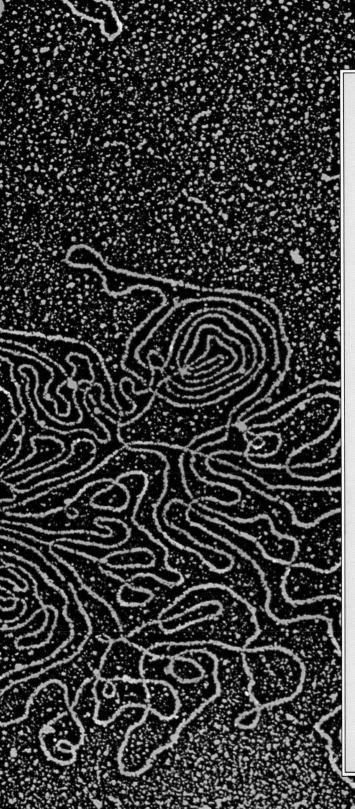

Mechanisms of Evolution

EVOLUTION is about changes and relationships. The living world does not remain static, but new species arise while others vanish. The new species are demonstrably related to older species. How does this come about?

We know that members of a family resemble each other more closely than people chosen at random—there are inherited characteristics. Yet children are not exact copies of their parents—variation also takes place. The key to inheritance was found by Gregor Mendel. His investigations of measurable characteristics showed that—in modern terms—each characteristic is determined by genes that are passed on from parents to offspring. We now know that the genetic material is a compound called DNA which is able both to replicate itself and to control protein synthesis very exactly.

The genetic make-up of individuals within a population varies: the same gene may have different forms producing variations on a given characteristic—and occasional mutations introduce completely new genetic forms. Natural selection acts upon this varied population, so that in time the gene frequencies alter. Normally, factors such as migration tend to mix the genes back in, and natural selection also tends to maintain species stability in large populations. How then can new species form?

Entire species probably do not often change into new ones. Sometimes it is in small populations, either geographically or ecologically isolated, that speciation takes place: genetic variation and selection have a much greater effect in producing change within a small community. This can be studied experimentally in short-lived species such as fruit flies.

◄ **The basic characteristics** and functions of a living organism are controlled by reproductive material called DNA. Normally it exists as long strands, which are found in every living cell. This electron-microscope photograph shows the total amount of DNA required for the functioning of a single bacterium (*Escherichia coli*).

VARIATION

The development of modern genetic theory. . . Genetic variation in a snail. . . The structure of the cell. . . Chromosomes and genes. . . DNA, RNA and the protein chain. . . How genes affect populations. . . Alcohol susceptibility in a fruit fly. . . Mutation, migration and random genetic drift. . . Different types of natural selection. . . Animals that mimic others. . . Natural selection in land snails and moths. . . Sickle-cell anemia. . . Human blood groups. . . Clines and gene flow. . . The genetic structure of island populations

DURING the mid-19th century, the main stumbling block to a better understanding of evolution was a complete ignorance of the mechanism of inheritance. It was clearly recognized that individuals who are related resemble one another more closely than random individuals, but the reasons for this lay surrounded in mystery.

The Determination of Variation

This ignorance was one of the factors which led to the development of a theory that traits acquired by an individual during its lifetime might be transmitted to the offspring. The theory is traditionally credited to the French biologist Jean-Baptiste de Lamarck (1744–1829), and has had several vigorous advocates down the years. The idea is an attractive one. If an individual such as a blacksmith develops big and powerful muscles through constant use, it seems reasonable that these traits should be transmitted to his sons. Observations of the powerful frames of blacksmiths' families were familiar in the 19th century, and lent support to the theory.

Similarly, the ancestors of giraffes might have stretched and elongated their necks by reaching up to the upper branches of trees. A long neck, developed by constant use, might have been passed on to the offspring, thus accounting for the evolution of the long neck of giraffes seen today. This theory could be applied to the evolution of any characteristic that played an important part in the life of an animal: the trunk of an elephant, the spade-like feet of a mole, the large ears of the Bat-eared fox, and so on.

Over the years, a variety of experiments have been performed that purport to demonstrate this method of inheritance, ranging from naively examining spaniels to see whether constant tail-docking results in a shortening of tail-length at birth, to the sophisticated breeding of animals in altered environmental conditions. The value of the former has been somewhat acidly likened to analyzing the inheritance of a wooden leg. The latter experiments are more difficult to assess. However, all of them seem either to have been poorly designed so that alternative explanations are possible, or to be incapable of repetition in other laboratories. Consequently, attractive as the theory may seem, it is not currently supported by any modern rigorous scientific study.

Charles Darwin was all too aware that a lack of knowledge of the mechanism of inheritance imposed severe limitations on his theory of evolution by natural selection. He attempted to circumvent the problem by inventing the theory of pangenesis, which was a modification of an Ancient Greek idea. He postulated that all cells in the body produced minute particles called gemmules, which circulated freely in the bloodstream and accumulated in the reproductive cells before being transmitted to the progeny. This was sufficient to explain his own observations on inheritance, and was compatible with the inheritance of acquired characters since every organ was said to produce its own gemmules. However, it was impossible to demonstrate experimentally, and had been abandoned long before the rediscovery of Mendelian genetics.

Mendelian Genetics

The breakthrough came with the publication in 1866 of the experiments and ideas of an Augustinian friar named Gregor Mendel (1822–84), which unfortunately lay unrecognized for 35 years until their rediscovery early in the 20th century. Mendel lived and worked in an abbey in Czechoslovakia, in a town which is now called Brno. His genius was to choose simple, easily quantifiable characters for study, rather than the more diffuse traits such as intelligence which had previously been examined.

The experiments for which he is most famous were performed using garden peas. This plant has both male and female organs in its flowers, but by selectively removing the pollen-bearing stamens (male organs) from some individuals it is possible to control the parentage in any particular experiment. Mendel began with a series of pure-breeding strains, such as those with tall or dwarf habit, white or green flowers, smooth or wrinkled seeds. He then used his emasculation technique to cross different strains, carefully counting the progeny. Some of the offspring he crossed among themselves, others he crossed back to the pure-breeding parental stocks. But he counted absolutely everything, and this quantification allowed regular and repeatable patterns to emerge, from which his theories developed—theories which in general have stood the test of time. Modern methods have produced a greater understanding of the mechanisms involved, but the basic principles still hold.

Modern Genetics

In contemporary parlance, the characters that Mendel studied are controlled by genes. All individuals possess two copies of each gene, one inherited from each parent. One copy of each gene is in turn given to every offspring, so that the hereditary material is transmitted, effectively unchanged, from generation to generation.

The variation that is apparent between individuals is due to the fact that not all copies of a particular gene are the same: they may exist in different forms (alleles, allelomorphs or allelic forms). For example, the land snail *Cepaea nemoralis* occurs in three different color varieties. These are produced by three different alleles of the shell-color gene. Every snail possesses two copies of this gene, which are either of the same allele, in which case the animal is said to be homozygous, or of different alleles, making it heterozygous. In the case of the heterozygote a complication arises in that the character determined by one or other gene may not be manifested. One allele often masks the effect of the other, in which case the character it determines is said to have been inherited in a dominant fashion. A character

▶ **Mendelian genetics.** The modern study of heredity is based on an understanding made in the 1850s and 1860s by an obscure Moravian friar, Gregor Mendel (1822–84). His research showed that contemporary assumptions about heredity were incorrect, and it provided a new basis for further study.

Mendel demonstrated that offspring inherit the factors for specific characteristics from their parents. The contributions from each parent are not blended, as was thought in Mendel's time, but some are exhibited in each generation whilst other are carried anonymously and transmitted to the next generation. Within each generation the characteristics appear in certain proportions.

An offspring inherits genes for a range of characteristics from both parents. In a brilliantly simple experiment Mendel studied the transmission of height. He crossed a pure-breeding tall pea plant with a pure-breeding dwarf pea plant RIGHT. All the offspring in the first generation (F_1) were tall, from which Mendel concluded that the inherited character for tallness (T) was dominant over that for dwarfness (t). He then crossed two plants from this generation and produced a second generation (F_2) in which there were three tall plants to every dwarf, showing that the factors for dwarfness must have been transmitted by tall plants.

In modern terms the original pure-breeding plants, each having similar genes for height, were homozygous. Descendants having different height genes are called heterozygous. In a heterozygous individual the characteristic that appears in the individual is said to be dominant, and the latent trait recessive.

In a further experiment FAR RIGHT Mendel crossed pea plants homozygous for two different characteristics. The varieties produced in the second generation showed that characters were combined independently rather than in groups. For example, a second-generation pea inherited yellowness from a yellow, round grandparent and wrinkledness from a green wrinkled grandparent.

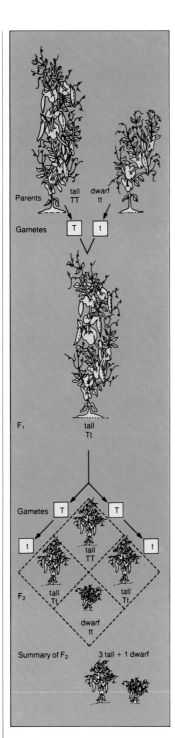

▶ **Friar Mendel at work** in his garden at the Augustinian house in Brno. Mendel's experiments with peas were conducted over seven years (1856–63) and required the establishment of pure-breeding strains.

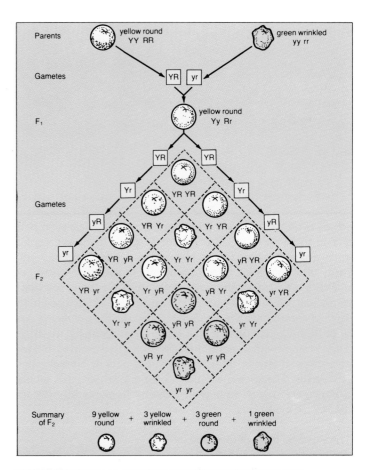

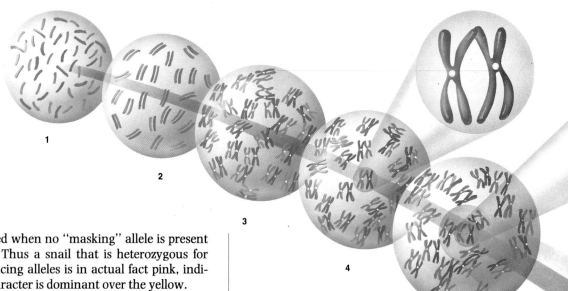

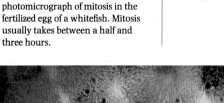

which is only manifested when no "masking" allele is present is said to be recessive. Thus a snail that is heterozygous for pink- and yellow-producing alleles is in actual fact pink, indicating that the pink character is dominant over the yellow.

This leads to an additional complication, for the appearance of an individual need not tally with its genetic composition. The appearance of an individual is called the phenotype, as distinct from its genotype, which relates to its genetic structure. Thus a snail with a pink phenotype may have either of two genotypes: it may be either homozygous pink or heterozygous pink/yellow. Recessiveness also accounts for those cases where a character that is apparently absent from parents occurs among their offspring.

The Physical Basis of Inheritance

At the moment of fertilization, a new individual consists of a single cell made up of an outer cell wall and inner cytoplasm, the main substance of the cell; this in turn contains the cell nucleus plus a number of other elements such as mitochondria and chloroplasts. The cell begins to grow and then to divide. First the nucleus divides into two daughter nuclei, and then a new cell wall is formed which partitions the cytoplasm and its contents into two halves. This process of cell-growth and cell-division continues until a mass of cells is formed. These

▼ **Mitosis underway,** a photomicrograph of mitosis in the fertilized egg of a whitefish. Mitosis usually takes between a half and three hours.

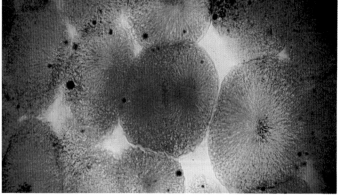

Mitosis

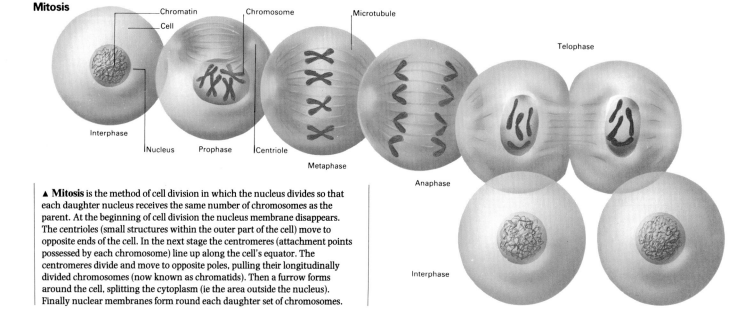

▲ **Mitosis** is the method of cell division in which the nucleus divides so that each daughter nucleus receives the same number of chromosomes as the parent. At the beginning of cell division the nucleus membrane disappears. The centrioles (small structures within the outer part of the cell) move to opposite ends of the cell. In the next stage the centromeres (attachment points possessed by each chromosome) line up along the cell's equator. The centromeres divide and move to opposite poles, pulling their longitudinally divided chromosomes (now known as chromatids). Then a furrow forms around the cell, splitting the cytoplasm (ie the area outside the nucleus). Finally nuclear membranes form round each daughter set of chromosomes.

◄▼ **Meiosis** takes place in the formation of sex cells (sperm and eggs), two of which may combine to form a new being (**1**). It differs from other forms of cell division in that the sex cells (or gametes) formed contain only half the normal number of chromosomes. (**2**) The chromosomes come together in pairs. (One from each pair was originally inherited from the individual's father and the other from the mother.) (**3**) Each chromosome shortens, thickens and splits lengthwise to form two chromatids. (**4**) The paired chromatids line up together in the dividing cell. (**5**) Lengths containing genetic information now "cross over," exchanging material between the maternal and paternal chromosomes. This is a crucial stage, at which genetic information from the two parents is reassorted. (**6**) The chromosomes separate, one member of each pair going to opposite poles. (For clarity stages 6–9 do not show the full number of chromosomes.) (**7**) The cell now begins to divide. Each part contains only half the original number of chromosomes, though each is split into two chromatids. (**8**) After the cell has divided completely each cell still contains the same number of chromosomes, each of which is doubled. (**9**) In a second division the chromosomes split in half and the halves are pulled to opposite poles. The cell then divides again. (**10**) Meiosis is now complete: each cell has a new mix of chromosomes, combining the parental traits in a different way from before.

cells then begin to develop differently into the various organs and tissues.

It is now known that almost every cell in an animal's body contains a complete set of genetic material. For this to happen, the genes must be able to replicate themselves before the cell begins to divide. This replication and separation of the genes must be extraordinarily precise if a single copy of every gene is to make its way into each daughter cell.

It was soon realized that there are particles within the nucleus which behave in exactly this fashion. These structures are called chromosomes, and can sometimes be seen under a microscope as rod- or thread-shaped bodies. There is usually an even number of chromosomes in the nucleus of a cell, and close examination reveals that they exist as a series of matched or homologous pairs, with the exception of one pair—the so-called sex chromosomes. The total number of chromosomes in a normal animal body cell is known as the diploid number.

The process whereby the chromosomes replicate and divide in normal body cells is called mitosis. A short time before cell division, the chromosomes replicate themselves, but remain attached at a point called the centromere, which need not be at the physical center of the chromosome. They then migrate to the center of the nucleus, and become aligned so that the daughter chromosomes lie on either side of a central plane. The centromeres then separate, and the daughter chromosomes disperse in opposite directions to leave a complete chromosome set in each daughter nucleus.

The formation of sex cells, or gametes, is basically similar to that of other body cells apart from one fundamental difference: every main body cell contains two copies of each gene, but only one of these is transmitted to each offspring. Thus the genetic material must be reduced by precisely half during gamete production. The complement of chromosomes within a gamete is known as the haploid number, which is exactly half the diploid number.

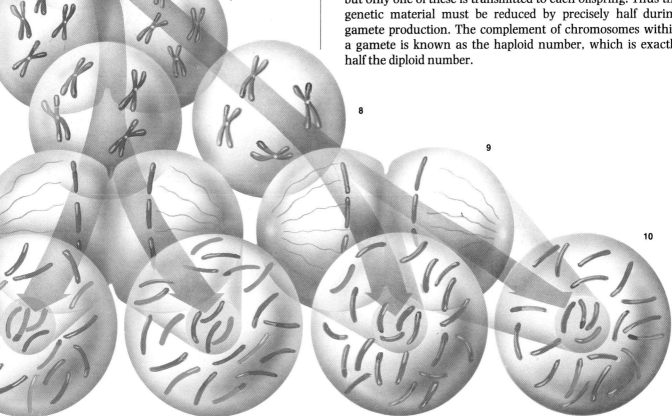

The process whereby sex cells divide is known as meiosis, and is initially similar to mitosis in that every chromosome replicates but remains attached at the centromere. This time, however, the chromosomes align themselves to lie in homologous pairs. The first division causes these homologous pairs to become separated so that one representative of each migrates to each daughter nucleus. A second division follows which is more like mitosis: the centromeres separate, and one copy of each daughter chromosome migrates to each of the new nuclei. This results in the production of gametes containing the haploid number of chromosomes.

These processes are basically similar to the behavior postulated for genes from breeding results. However, there are always far more genes than chromosomes, so the two cannot be directly equated, and it was suggested that the genes lay in or on the chromosomes. Support for this came in the early days of genetics, when it was found that some genes are not inherited independently but appear to be associated or linked to each other. These associations led to the discovery that genes fall into specific linkage groups, and that the number of linkage groups in a species is in general the same as the haploid number of homologous chromosome pairs. Thus the fruit fly *Drosophila melanogaster* has four linkage groups, and cytological examination reveals four pairs of chromosomes.

Although linked genes are carried on the same chromosome and tend to be inherited together, the linkage is not absolute. In breeding experiments offspring sometimes have gene combinations which are present in neither parent. This is due to a process known as recombination, which accords well with the observations of early cytologists, who noticed that early in meiosis homologous chromosomes sometimes appeared to be attached so as to form an X-shaped connection called a chiasma. Now chromosomes are usually long, filamentous structures that can easily break and hence need to be repaired; there is a series of enzymes that carry out this repair by locating the broken ends of the two fragments and bringing them together before reattaching them at the point of cleavage. It is now known that the two processes of recombination and chromosome repair are closely similar. If chromosomes break when they are closely opposed in homologous pairs, it is easy for the repair enzymes to locate the ends of the "wrong" fragments. This means that part of one parent chromosome will "cross over" to join the remainder of the other, with the result that the genes are recombined. The farther apart the two genes lie, the greater the chance of recombination—a factor which has been used to produce chromosomal "maps" for commonly investigated species such as fruit flies, mice, chickens, wheat, the bacterium *Escherichia coli*, and even man.

The Genetic Code

The years from 1950 onwards saw the development of the science of molecular biology, of which one of the most spectacular achievements was the elucidation of the molecular structure of genes and chromosomes. It is now established that the genetic material incorporated into the chromosomes consists of a long slender molecule called deoxyribonucleic acid or DNA. Although this molecule is very long, its structure is relatively simple, consisting of two strands each composed of a trio of units which is repeated tens of thousands of times.

Each strand consists of a linear arrangement of molecules of the sugar deoxyribose connected via phosphate radicals. This forms the backbone of the strand, and to each sugar molecule a base is attached, which can be of any one of four types: guanine, cytosine, adenine or thymine. The unit consisting of sugar, phosphate and base is called a nucleotide, and each strand thus consists of a string of polymerized nucleotides, or a polynucleotide. The two strands are held together by special links called hydrogen bonds, and the method of attachment is very precise. Bonding takes place between the bases, and only two combinations of base pairs are possible: guanine with cytosine, and adenine with thymine.

When this structure was deduced, it was quickly realized that it gave a simple yet accurate method for gene replication. If the molecule splits longitudinally, by the breaking of the inter-base hydrogen bonds, each strand can act as a template for the synthesis of a new complementary copy. Bases, sugars and phosphates are assembled into nucleotides, which are selected to complement the next nucleotide in the parental strand.

Mutations

The result is an extremely accurate method for gene replication during meiosis and mitosis. But the rare errors known as mutations can also be explained. Mutations occur when there

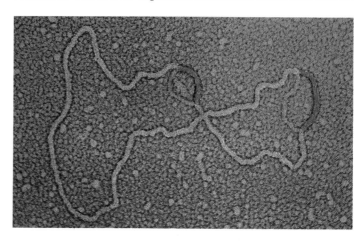

▲ ▶ **Genes and chromosomes.** To understand the nature of genetic mechanisms Mendel and his successors had to rely on deductions from experimental data. However it is now possible to determine the locations of particular genes on chromosomes, and this has been done for many animal species. In 1934 it was discovered that the chromosomes in the salivary glands of fruit flies in the genus *Drosophila* (1) are so large that considerable detail can be seen on them (4). As this fruit fly breeds rapidly it was possible to correlate variations in the flies (2) with changes in bands on the chromosomes, and eventually to produce a "map" (3) showing which sections of the chromosomes are responsible for particular features.

It is now also possible to show on photographs the areas on chromosomes or DNA strands occupied by particular sets of genetic information. The picture ABOVE is a false-color electron micrograph of a bacterial DNA plasmid (ie a length of "free-floating" DNA). The areas in blue are those carrying two genes. We have no reason to doubt that the genes of higher animals look any different.

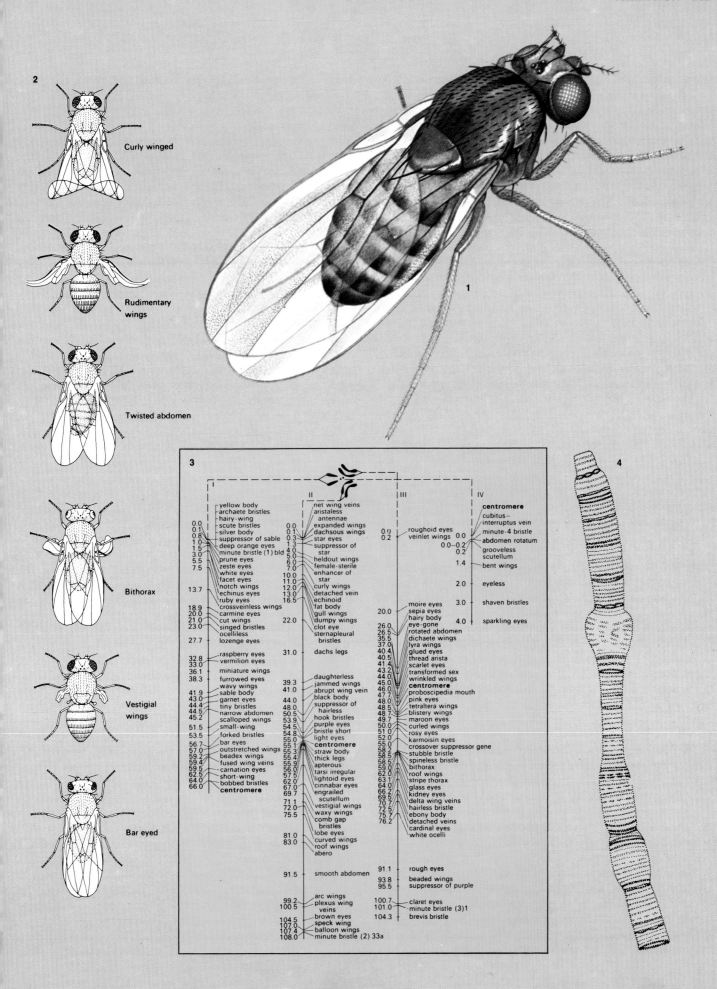

is an error in the replication of the genetic material, and may be due to a variety of causes. At a molecular level, however, mutation can happen when the "wrong" nucleotide is inserted into the new strand. Alternatively, one or more nucleotides may be lost, either by deletion before replication commences, or by the omission of one of them during the replication process. Thirdly, extra nucleotides may be inserted into the new DNA strand during the replication process. All of these will result in the production of a mutated gene, and—provided that this is not lethal—the incorrect sequence will be faithfully copied during the subsequent replication processes.

Transcription and Translation

The information held in DNA is converted into protein in a two-step process. The first stage is called transcription, which works as follows. First the DNA complex is separated into its two constituent strands. Only one of these strands is used in protein synthesis, acting as a template for the production of an intermediate molecule called messenger ribonucleic acid or mRNA. RNA has essentially the same composition as DNA, except that it usually exists as a single strand, has ribose as a sugar instead of deoxyribose, and has uracil as a base instead of thymine. Enzymes called RNA-polymerase locate the beginning of the gene in the DNA, and move along it, assembling a complementary strand of RNA according to the base sequence of the DNA. When the RNA-polymerase reaches the end of the gene, transcription is terminated, and the mRNA molecule moves out of the nucleus into the cytoplasm for the second stage of the process, which is called translation.

Proteins are composed of chains of amino acids called polypeptides. In translation, the sequence of amino acids from which a particular polypeptide is assembled is ultimately determined by a sequence of codes in the mRNA known as the codons, each of which is made up of a set of three bases. Present in the cytoplasm are a group of smaller RNA molecules called transfer RNA or tRNA molecules. Each tRNA molecule contains just one corresponding, complementary set of three bases called an anticodon. Each codon in the mRNA pairs up with the tRNA molecule whose anticodon exactly matches it. Each tRNA then binds to the amino acid which corresponds to its anticodon. Thus the sequence of amino acids in the final protein depends on the sequence of tRNA molecules, which in turn is determined by the sequence of codons in the mRNA that was transcribed from the original DNA sequence.

The discovery of this precise correspondence between DNA, RNA and the amino-acid sequence was one of the great biological discoveries of all time. It explains not only how hereditary information is passed virtually unchanged from cell to cell and from parent to offspring, but also how identical protein molecules can be repeatedly synthesized in the same or in different individuals.

It has now become evident, however, that this is not the whole story. There are sequences of DNA that intervene between functional genes and appear not to be used as genes. There is also evidence that genes can sometimes contain base sequences that have no corresponding amino-acid sequence in the finished protein. These are presumably excised in some editing procedure during translation. Nevertheless, such excep-

tions are merely refinements, and do not affect our basic understanding.

Genes in Populations

That individuals vary has always been known, but the study of this variation in nature has traditionally fallen within the domain of ecology and related disciplines. In recent years it has been realized that much of this variation is genetically rather than environmentally determined. Consequently geneticists are now moving into the field, bringing new methods and new insights to complement the researches of ecologists.

With a knowledge of Mendelian genetics it is possible to examine variation at several different levels. One can analyse the genetic structure of individuals, and identify them as homozygotes or heterozygotes. One can then screen whole populations, record the different physical forms (phenotypes) and estimate the frequencies of different alleles. The results of such research provide remarkable insights into the genetics not only of individuals but also of populations, of subspecies and even of whole species. The application of genetic analysis to whole populations with reference to their physical and biological environment has led to the development of the new disciplines of population genetics and ecological genetics.

It is possible to find correlations between allele frequencies in populations and the different ecological constraints which act upon them, and this would suggest that the function of genes is in some way influenced by environmental constraints.

For example, in the fruit fly *Drosophila melanogaster* there is a gene that produces an enzyme called alcohol-dehydrogenase (*Adh*). This substance acts by removing a hydrogen atom from alcohol molecules. In the wild, the adult *D. melanogaster* lays its eggs on fallen fruit; the larvae burrow into the fruit, feeding on its tissues and on the yeasts that grow there. Among the by-products of yeast metabolism are alcohols, some of which may be present at levels which are toxic to the growing larvae. *Adh* acts here as a detoxifying agent, and may thus be very important to the survival of individual larvae.

The gene which codes for *Adh* has two forms (alleles), each producing a subtly different form of the enzyme. The difference between these two forms may seem trivial, but ecological geneticists have found striking differences in the distribution of the two alleles. All other factors being equal, one allele is commoner than the other at high altitudes and in more northerly latitudes. It also appears to occur more frequently in sites with a high incidence of environmental alcohol, such as vineyards and wine cellars. Consistent patterns like these imply systematic causes, and are meat and drink to the ecological geneticist.

Why do the two alleles of the *Adh* gene differ so greatly in their distribution patterns? The answer lies partly in the biochemical properties of the two resulting forms of the *Adh* enzyme. Although they differ in only a single amino acid, this has a dramatic effect on their biochemistry. The first form has a much higher reaction rate, and so eliminates potentially lethal alcohol molecules from the larval tissue more rapidly than the second. On the other hand, the second form remains stable up to much higher temperatures than the first.

The relative frequencies of the two alleles in a population would therefore appear to depend upon the temperature and

▼ **The basis of mutation.** The features of an animal take their expression ultimately from a code embodied in each cell's genetic material (DNA). This code specifies the order in which amino acids are to be formed into proteins (within the area outside the cell) which in turn control the cell's structure and metabolism. "Mistakes" in the genetic code can therefore have far-reaching effects.

Within the DNA are four substances (nucleotides). These are lined up in combinations of threes (triplets) which specify (via Messenger RNA) that particular amino acids should be included in a cell's synthesis of a protein. Shown here (1) is a sequence of the triplet formed by the nucleotides cytosine (C), guanine (G) and adenine (A), which specifies the amino acid alanine.

A genetic mutation occurs when there is a change in a triplet within a sequence, which may make that triplet and possibly others adjacent code for a different amino acid. The simplest change (2, substitution) affects only one triplet and codes for a single different acid. The deletion (3) or insertion (4) of a nucleotide alters many triplets within a sequence, having a "knock-on" effect. The outcome of such mutations can be dramatic and debilitating.

Mutations occur fairly frequently during the replication of genetic material (eg during meiosis, the production of sex cells), but comparatively few survive to produce full-grown organisms. Mutations are, however, one means by which variety is generated during evolution.

DNA and Protein Synthesis in Cells

Proteins (ie compounds essential to living organisms) are actually "built" within the cytoplasm (the outer area of a cell) on small bodies called ribosomes, but their synthesis is controlled from within the cell nucleus by the DNA of the chromosomes, with molecules of ribonucleic acid (RNA) acting as the messengers between the nucleus, cytoplasm and ribosome.

(1) The sequence of amino acids which make up each individual protein molecule is coded in the DNA molecule. The code for each amino acid is held in the sequence of the pairs of complementary bases (thymine, adenine, cytosine and guanine) which link the twin helices of the DNA molecule. A sequence of three nucleotides on the DNA molecule, known as a codon, will code for a single amino acid. The sequence of codons that codes for all the amino acids in a polypeptide chain can be called a gene. (2) The DNA helix uncurls and a chain of messenger RNA (mRNA) is synthesized with complementary nucleotides to these on one of the DNA strands; the mRNA thus contains all the codons required to form a single polypeptide chain. (3) This mRNA molecule then makes its way through a pore in the nuclear membrane into the cytoplasm where it attaches to a ribosome (4). (5) Meanwhile another RNA molecule called transfer RNA (tRNA) is synthesized in the nucleus and makes its way out into the cytoplasm (6). Such tRNA molecules are specific for each amino acid, having the ability to "pick up" these specific amino acids from the pool of amino acids found in the cytoplasm. (7) The tRNA molecules plus their amino acids then migrate to the ribosome. (8) Each tRNA molecule also possesses the complementary codon for the codons on the mRNA molecule, so that these complementary codons join at the ribosome. (9) The bond between the two RNAs breaks down, allowing further tRNA, plus further amino acids, to link to the succeeding mRNA codons. (10) In this way the chain is progressively lengthened.

This description of the process does not do justice to its speed: it takes only one second to form a chain of 4,000 acids.

Thymine (uracil in RNA)	
Cytosine	
Adenine	
Guanine	

A	C G A	C G A	C G A	C G A	C		
	Alanine	Alanine	Alanine	Alanine		1	Normal

A	C G A	C C A	C G A	C G A	C		
	Alanine	Glycine	Alanine	Alanine		2	Substitution

A	C G A	C A C	G A C	G A C	G		
	Alanine	Valine	Leucine	Leucine		3	Deletion

A	C G A	C A G	A C G	A C G	A		
	Alanine	Valine	Cysteine	Cysteine		4	Insertion

the alcohol content of the environment. Prolonged high temperatures will destabilize the more active form of the Adh, thus reducing the fly's chances of survival. High alcohol concentrations will be detrimental to flies possessing the heat-stable, but less active Adh, thus reducing the frequency of the associated allele in the breeding population. The relative frequencies of the two alleles thus depend upon a complex environmental interplay, coupled with other effects that are not properly understood.

The situation where two or more alleles are present in a population is known as genetic polymorphism. (This term is not applied when the alleles concerned are very rare in a population or have been introduced by means of a mutation or a very rare immigrant.) A situation of genetic polymorphism is subject to a large number of complex interacting factors, as is shown in the case of alcohol-dehydrogenase.

However, when the effects of natural selection are reduced to a minimum, a number of regular patterns emerge—so regular that they can be reduced to mathematical formulae. For example, the proportion of homozygotes in such a population is equal to the square of the allele frequency, while the frequency of heterozygotes is equal to twice the product of the frequencies of their constituent alleles. This rule is one of the cornerstones of theoretical population genetics, and is called the Hardy–Weinberg Law after the two men who discovered it independently in the early years of the 20th century.

Factors Affecting Variation

One of the advantages of Mendelian genetics is the marvelous basis which it provides for understanding some of the mechanisms of evolution. It is easy to see how evolutionary change can be brought about by the substitution of one allele for another within a population, or even by a change in gene frequencies. This is not to say that all evolution comes about in this way, but an increase or a decrease in allele frequency, for whatever reason, will bring about a change in the genetic structure of a population or species.

Perhaps the most obvious factor influencing gene frequencies is that of natural selection, as illustrated by the fruit-fly example above; but there are a number of other factors which should also be considered.

The first of these is mutation. It is inevitable that mutations will occur in populations, and unless they are lethal or cause infertility, they will be transmitted from parent to offspring. This will result in a decrease in the frequency of the unmutated allele and an increase in that of the newly formed one. It is now generally agreed that mutation occurs only about once in every million replications. This means that new alleles are produced at an extremely low rate—far too low, in fact, to be of any significance on their own in changing the frequency of an allele in a whole population. However, mutation is of vital importance in that it is the only means by which truly novel genetic material can be produced. Thus its role is simply to produce genetic variants, while any changes in the actual frequency of the resulting variants will depend on other factors within the population.

A potentially more important factor in changing allele frequency is a phenomenon known as random genetic drift. This tends to occur in a population as a result of random differences in the survival of different alleles. For a population to remain genetically stable, allele frequencies must remain exactly the same from generation to generation. This is never likely to be completely true, as chance will inevitably produce random differences in the survival rates of different groups within the population, and the allele frequencies will differ slightly as a result. In the absence of any restoring factors, these marginal changes will be passed on to the offspring. Thus allele frequencies will tend to drift randomly up or down.

Random genetic drift will only have a significant effect in small populations, where accidents to a few offspring will have more appreciable effects on overall allele frequency. Its effects will also be more pronounced in divided or isolated populations,

▼▶ **Genetic variation** between individuals of a particular species can often be seen in coloration and other prominent external features. BELOW varieties of the snail *Cepaea nemoralis*. RIGHT dark and normal forms of the European adder (*Vipera berus*). The frequency of particular varieties can often be shown to be determined by natural selection.

▶ **Gene frequencies in populations.** Mendel discovered that the frequencies with which the factors controlling a feature are expressed and transmitted within a family can be reduced to a statistical formula. This principle was later extended to gene frequencies within large populations.

If the individuals in a large population mate at random, and if a gene locus has two alleles A, a, then the following combinations will be formed

Father's gametes

		A	a
Mother's gametes	A	AA	Aa
	a	Aa	aa

If the frequencies of A and a are denoted by p and q then the frequencies of new individuals will be as follows:

	A p	a q
A p	AA p^2	Aa pq
a q	Aa pq	aa q^2

So a total population will consist of:
$p^2 + 2pq + q^2$.

The constancy of the relationship between the production of gametes and the proportions of the genotypes of new individuals is exemplified by sea anemones. Adult anemones shed gametes into the water in the same proportions as they occur in the adults. These gametes unite randomly to produce larvae which in turn settle down onto rocks where they grow to be the next generation of adults. If the population is large and breeds randomly the *proportion* of each genotype (ie the genetic constitution) in the progeny population is equal to the product of the constituent genes.

The importance of this formula is that it permits the calculation of expected gene frequencies in a population against which change brought about by natural selection, mutation, migration or chance sampling can then be measured. It was worked out separately in 1908 by W. Weinberg, a German physician, and G. H. Hardy, an English mathematician, and is called the Hardy–Weinberg Law.

since the frequency of a particular allele may go up in one locality while it goes down in another, gradually causing the two populations to diverge. This has been suggested as a cause of the large differences in blood-group frequencies in the small, disparate village populations in the South American jungle.

A factor which militates against random genetic drift is that of migration. Divergence between adjacent populations, whether as a result of random genetic drift, natural selection or whatever, will be retarded if there is movement of individuals between the two populations. In cases where there is little or no effect from natural selection, genetic drift can be overcome by as little as one migrant per generation, thus preventing the divergence of the populations concerned.

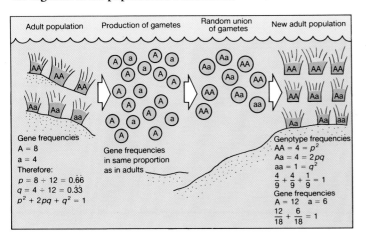

Natural Selection

Despite the various effects of mutation, migration and random genetic drift, the prime factor affecting evolution is that of natural selection. Darwin and Wallace were not aware of Mendelian genetics, but it is relatively straightforward to interpret their ideas in terms of genes and alleles.

Variation is widely seen in nature, and it may be genetic or environmental in origin. In some situations it is possible to calculate the proportion of phenotypic variance that is genetic in origin. This factor is known as heritability, and a low value implies that much of the variation is due to environmental causes. Attributes which have a high heritability in man include height and blood pressure, while those with a low heritability include susceptibility to measles and, perhaps surprisingly, weight.

In every species there is a substantial overproduction of offspring in each generation, and this means that a high proportion of individuals will die before they can produce offspring. Those individuals with the attributes necessary to win this struggle will survive and reproduce, thus transmitting their genes to the succeeding generation, and passing on those attributes that are heritable. This directional form of selection has been inferred in the evolutionary lineages of a variety of groups for which fossil records are available. It is, however, easier to see in artificial situations. For example, the dramatic differences in size between various breeds of horses can be attributed to deliberate and strong selection over many generations.

▲ ◄ **Selective breeding** is one means by which variation can be induced, by man. Its force is demonstrated by the extreme sizes of breeds of some animals, for example horses: ABOVE Shire horses, LEFT Falabellas. Excessive inbreeding, however, can upset the balance of an animal's physique. The Falabella has lost much of its physical strength. Some specially bred pets are extreme distortions of their kind.

However, selection does not always act directionally. It has, for example, been shown that when a population has reached an optimal standard, selection tends to eliminate those individuals that depart most from it. A good example of this is the weight of human babies at birth. The majority are born weighing close to 3kg (7lb), and mortality is low among these. Mortality is severely increased in babies whose weight is significantly higher or lower than this "optimal" value. This is an example of stabilizing selection: the elimination of those individuals that are extreme, whether for genetic or for environmental reasons.

There is a converse to this. Disruptive selection tends to favor individuals that possess extreme phenotypes. This form of selection is not common in nature, but there is a case of mimicry which is a particularly good example. Females of the Asian swallowtail butterfly *Papilio memnon* exist in about 20 different forms. All of them are palatable to predators, but many are mimics of certain unpalatable butterflies with which they coexist. The color and pattern of the mimics is controlled by a series of five closely-linked genes, each with allelic variants. The various mimetic forms are produced by different combinations of these alleles; there are perhaps two or three of these combinations in any particular location, and intermediate forms are exceedingly rare. Mimetic animals depend for their survival on an exact resemblance to one or other of the dangerous models. Inexact matching reveals to the acute predator that they are different from any dangerous model and so possibly palatable. Selection thus tends to favor those "optimal" phenotypes that most closely resemble the unpalatable butterflies. The problem has been resolved in *P. memnon* by the evolution of a series of closely-linked genes which are inherited together as a kind of "supergene." Any individual that recombines will be intermediate in appearance, resemble none of the dangerous models and be killed; but the closeness of the genetic linkage means that such an individual is only rarely produced.

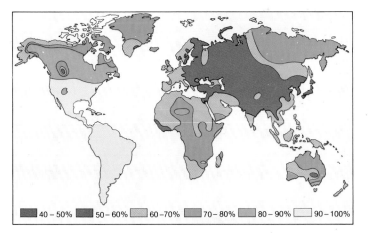

40 – 50%   50 – 60%   60 – 70%   70 – 80%   80 – 90%   90 – 100%

▲ **Genetic distinctions between races.** There are four common ABO blood groups in humans. The group to which each person belongs is determined by genes, but the relative distribution of blood groups varies throughout the world; different races have different blood-group frequencies. The map shows the frequency of the blood group O in aboriginal populations of the world.

The factors determining selective advantage or disadvantage are thought by some to be constant and unchanging. However, this need not always be the case, and there are those who would say it is never the case. The relative advantages of two phenotypes can vary depending on their relative frequency and upon how abundant they become. For example, if a mimetic form becomes too abundant, predators will learn to associate its appearance with a palatable rather than with a noxious item. Similarly, the relative advantages of two mimetic forms can be influenced by their relative frequencies: if one form is particularly common, then it is likely to suffer greater predation than a scarcer mimic.

The Geographical Distribution of Genes

Ecological genetics is the branch of science that is concerned with the effects of genetic variation upon the ecology and population biology of animals and plants. One important area of research has been the study of allelic distributions in natural populations. These are in fact extremely variable: alleles can be abundant in some populations and scarce in others. Initially it was thought that this might simply be a matter of random differences between divided populations, but studies in a variety of species have revealed patterns that are too consistent to be attributed to random effects.

For example, many woodland populations of land snails such as *Cepaea nemoralis* and *Arianta arbustorum* include high frequencies of alleles that produce dark phenotypes. Adjacent populations in open habitats such as grasslands have higher frequencies of alleles producing pale shells. Similarly, in many species of moth with different-colored forms, the dark phenotypes are more abundant in urban environments, while pale forms predominate in the country. These patterns can be explained on the basis that dark or light forms will predominate in the areas where they are better camouflaged against predators. Indeed, in several instances bird predation has proved to be the main selective factor.

Consistent patterns of distribution that are associated with the environment are not restricted to visual traits. In our own species, for example, there is an allele which in the homozygous state (ie when inherited from both parents) produces sickle-cell anemia. Those suffering from the disease usually die young, and the allele is consequently very rare in many human populations. However, homozygous "normal" individuals (ie those in which no such allele is present) are susceptible to malaria. Thus in areas where malaria is endemic they too tend to be selectively eliminated. Here the balance of forces favors heterozygotes, who suffer from neither disease. Consequently the allele for sickle-cell anemia is much commoner in malarial regions.

The selective agents are not always so plainly evident. In the Mojave Desert in California, there is a small plant called *Linanthus parryae*, which has alleles producing two distinct flower colors. There are striking differences in the frequencies of blue and white flowers from place to place, and sometimes from year to year. These changes appear to occur independently of any environmental factor, and yet they are too significant to be attributed to random genetic drift. But despite many years of study, the factors which might explain the variations in this enigmatic little plant have remained shrouded in obscurity.

There are also profound differences between the human races in terms of blood-group frequencies. In some instances these have been ascribed to the selective effects upon blood groups of diseases that are prevalent in different regions. In other cases they remain unexplained. It may be that the scientists have not yet looked hard enough, or have simply missed the significant ecological or environmental factor. Alternatively, the causes of the differences may be buried deep in a complex of interlocking factors of which selection and random processes are merely a part.

Natural populations sometimes reveal patterns of allelic distribution that involve gradual change rather than sudden discontinuity. These progressive changes in a phenotype are known as clines.

A familiar example of a cline is the way that the ears of North American rabbits become progressively shorter with increasing latitude. The thin, vascular tissue of the ear is an area of potential heat loss—a factor which can be advantageous in hot climates and deleterious in cold ones. Rabbit populations have adapted to this, so that now there is a close correlation between ear area and environmental temperature.

Clines are not only found in visual or quantitative factors such as ear size, hair color or clutch size. In several cold-blooded species, clines have been detected in enzyme alleles that correlate with environmental temperature. Close analysis has shown that the warmer the climate, the more predominant the allele producing enzymes which are stable at high temperatures or which are more efficient when the body temperature is high. This is unsurprising for cold-blooded animals, as they are much more dependent on the temperature of their surroundings.

On the other hand, clines are not always due to natural selection. If two adjacent populations of a species are isolated for a long period, they may evolve differences either selectively or randomly. When they come back into contact, movement between the two populations will result in gene flow, and the sharp change in gene frequency at the boundary between them will become progressively more blurred, thus creating a cline. The mobility of the species concerned will also be critical, since the boundary between the two populations will tend to remain sharper in a more sedentary species such as a snail than in a more mobile animal such as a moth or a bird.

The Effect of Isolation on Population Genetics

The effect of isolation, and of barriers to gene flow, in encouraging evolutionary divergence and speciation is nowhere more apparent than on islands. The study of island populations has revealed consistent faunal patterns that can be directly attributed to their isolated situation.

Islands tend to have fewer species than adjacent land masses. Thus Ireland possesses only 60 percent of the birds that breed regularly in Britain. Small islands usually have even fewer species, and the species they do have are frequently more "general" in their ecology than related forms on larger islands. The limited number of species on islands is due not only to difficulties of dispersal, but also to ecological factors. Animals on a particular island become adapted to local conditions, and if these are relatively uniform, a small number of more generalized species will tend to thrive at the expense of a large number of specialists. In times of hardship, a generalist can switch its food supply and survive, whereas a specialist cannot and so dies.

Many of the animals found on islands are endemic, ie they occur nowhere else. Thus 21 percent of the birds of Iceland are endemic subspecies, as are 45 percent of those of the Azores. Groups of islands can reveal even more interesting patterns. Their animals and plants may show close similarities which are not shared with other more distant populations, but there are frequently slight differences between the populations of individual islands. For example, giant tortoises occur in the Galapagos Islands and nowhere else in South America or the Pacific Ocean; but they are not always similar, for the tortoise populations of different islands often show marked differences in the shape of the shell.

The barrier created by even a short stretch of sea means that ecological adaptations and genetic differences can accumulate on an island without the diluting effects of gene flow from large mainland populations. Hence the unique faunas that have evolved and developed on remote islands. However, the very smallness of the habitat makes it much more vulnerable to external forces, and there is evidence that extinction occurs much more frequently among isolated populations. This takes place in the natural state, but is often accelerated by human intervention. Direct and deliberate destruction of the habitat can of course be fatal. More insidious, perhaps, is the introduction of new predators to which the local fauna is not adapted, or of competitors that are more efficient and force them into less viable habitats. DTP

◄▲ **Island adaptability.** A classic example of how one species has evolved genetically into several different species is that of the Galapagos finches, a group of 13 species (belonging to four genera) found only on the Galapagos Islands. Body size is roughly the same in all of them, but they occur in a wide range of feeding niches which would be occupied by unrelated bird species in a mainland situation. This "adaptive radiation" is shown in their bills which differ in size and shape.

(1) A Large cactus finch (*Geospiza conirostris*); seed-eater. (2) Warbler finch (*Certhidae olivacea*); small insects from trees and shrubs. (3) Small ground finch (*Geospiza fulginosa*); ground-feeding seed-eater. (4) Cactus finch (*G. scandens*); arboreal seed-eater. (5) Large tree finch (*Camarhynchus psittacula*); arboreal insect-eater. (6) Woodpecker finch (*C. pallidus*); insects from trunks and branches. (7) Sharp-beaked ground finch (*Geospiza difficilis*); arboreal seedeater. (8) Large ground finch (*G. magnirostris*); ground-feeding seed-eater. (9) Vegetarian finch (*Platyspiza crassirostris*); buds and leaves.

THE ORIGIN OF SPECIES

*What is a species?... How does a new species form?...
Populations split up into separate areas or ecological
niches... What are the mechanisms of major evolutionary
changes?... The evolution of the Polar bear from the
Grizzly bear... Evolution goes in fits and starts... Studies
of fruit-fly species on Hawaii... Insects that have evolved
in response to industrial pollution... Insects develop
resistance to insecticides*

ANIMALS and plants alive at the present are the changed descendants of animals and plants that lived in the past. We
know this because the origins of both an individual animal and
plant and its immediate family are precisely known from direct
scientific observation. Fast-breeding forms, such as *Drosophila*
flies, can be traced back hundreds or thousands of generations.
The first mutant *Drosophila* stock (white eyes rather than the
usual red) was established about 1910. In this fly a generation
takes about 10 days so that this stock has been observed for
more than 2,200 generations.

This stresses the historical nature of the genetic material of
each species of plant and animal, but it is not possible to observe
directly the reproductive events that occurred in the generations that occurred in the remote past: there is no data available
to trace the genetic origins of species with the same precision
as more immediate ancestry.

The term species usually means one kind of animal or plant.

If origins are to be considered, however, it makes sense to view
a species as a series of local populations of different sizes and
geographical locations. Genetic studies show that there is
abundant genetic variation within a species. Each individual
member of a species, or even of a very small subdivision of a
species, like a family unit, is different from the next genetically.
This is true even though the variability may not be easily
apparent to direct visual observation. This genetic diversity,
clearly seen by the study of proteins and the DNA code itself,
is one of the key discoveries of modern population biology.

The species is a reproductive community of one kind of
organism that consists of many genetically variable populations in different places. Yet this community has a boundary.
The species does not blend into one or more other species: it
tends to be a reproductively isolated group. The phrase "tends
to be" is used since the reproductive isolation of a species from
a closely related one is often not absolute. The lion and tiger
are two distinct species of big cat (genus *Panthera*). They do
not occur naturally near one another and do not mate together.
Yet hybrids between them can be produced in captivity. Many
plant species will produce hybrids with members of a neighboring related species, and such hybrids are often fertile. In the
natural world, however, species rarely lose their genetic identity in a swarm of hybrids; zones of hybridization are usually
narrow and local.

So it is not easy to define what is meant by a species. The
best solution lies in emphasizing the nature of a species as a
group of organisms sharing a common gene pool and showing

adaptations that fit their members to a particular environment. If these adaptations become genetically very precise and exclusive (lion to grasslands, tiger to forest, for example) the species may show little or no hybridization, partly since no opportunity exists for interbreeding. On the other hand, if the species is more loosely adapted, some natural hybridization may occur, without the species losing its identity. In short the species is best viewed as a set of populations in which the limits of genetic variability are maintained by natural selection.

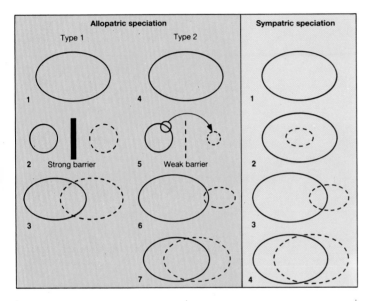

▲ The origin of new species. New species may arise because of some restriction or isolation in the range of a parental form. In allopatric speciation isolation is initially geographical; in sympatric speciation it is initially a more restricted confinement to an environmental niche.

There are two main models of allopatric speciation. In Type 1 a strong geographical barrier splits a population (1) into two large subpopulations (2). Each then evolves independently. If the barrier disappears or is surmounted once a degree of divergence has occurred the new species will not cross and will maintain their independence (3).

Sometimes a small subpopulation will become partially isolated (5), but the isolation is such that sufficient differentiation takes place so that if it is then reunited with the original population its distinctiveness will remain (6), even though its range may then spread over some or all of the area occupied by the original population (7).

In sympatric speciation individuals within a large population (1) occupy a particular local environmental niche and through inbreeding and natural selection become adapted to it (2). They become genetically distinct and eventually unable to cross successfully with other descendants of the original population. In time they may occupy similar niches across a larger range (3) and eventually be found across the entire area occupied by the original population (4).

◄ The route of no return. Although the genetic borders of species are not impervious, ie *species* that interbreed can sometimes produce offspring, it is not common for them to remerge completely with each other. This is a zebroid, the outcome of union between a zebra and a horse. These animals belong to the same family but their progeny is sterile. This is because there is insufficient compatibility between the genetic materials inherited from the parents to produce sex cells (gametes). More fertile hybridization occurs in the plant kingdom.

Speciation as a Process

Speciation means the processes by which a new species is formed from some preexisting population. All the schemes outlined here are hypothetical, since only present species and their populations can be observed.

Species clearly arise as genetically altered populations of preexisting species. But what are the precise conditions under which this can occur? The population biologist approaches this question in two ways. First, he may look among the existing related groups of animal or plant species for two or more that give evidence of having been formed only recently. By studying the genetic, ecological and geographical differences between them, he may be able to suggest how they came to be separate. A second approach is to seek out a single species and to look for evidence of the beginning of splitting; that is, are there one or more local populations that might be viewed as an incipient species?

Subspecies as Incipient Species

One major theory of speciation is based on the widely observed fact that many species show geographically distinct populations known as subspecies. In some cases these can be shown to be adapted to their local ecological conditions. Should this area be cut off from contact with the main body of the species, the isolated population might then begin to show intensified adaptation to local conditions. If geographical contact with the ancestral form is later renewed, the genetic differences between the new group and the old might be great enough to result in reproductive incompatibility or at least to result in only a narrow zone of hybrids. The hybrids may well be poorly adapted to both ecological situations, and this may reinforce the partial incompatibility through natural selection. In turn this may result in reproductive isolation of the new group from the old. This hypothetical process is usually referred to as allopatric or geographical speciation. Some authors feel that reinforcement by natural selection is an unlikely basis for the observed reproductive isolation. One alternative suggestion is that the new adaptations acquired in isolation may indirectly hinder interbreeding. Speciation, in this view, is largely an incidental effect of the genes that are subject to direct selection for their specific adaptive properties.

Another variant of the above scheme requires a drastic reduction in population size locally. This can occur when a population, peripheral to the main species, is established by one or a few founder individuals. The new population, by virtue of the fact that only a portion of the old gene pool was brought along with the founders will almost certainly have different gene frequencies from the ancestral populations and consequently may now be forced to reorganize its genetic material. In particular, the functioning of the reproductive system might be changed, again resulting in incompatibility with the old species.

Whatever its mode of origin, it is clear that speciation requires reproductive isolation, and many studies have been made of those features of two closely related species that appear to be responsible for such isolation. For example, there might be some genetically-determined incompatibility in mating behavior. In other cases, the mating process may apparently occur normally but offspring are either not produced at all or are sterile.

Many detailed observations of the natural history and geographical distribution of species strongly suggest that allopatric speciation has been widespread in nature. Certain groups such as *Drosophila* flies and mice are highly speciose, that is, many species have been formed in the past. Those that are most closely related often appear to replace one another geographically, suggesting that the differences observed between subspecific populations might be capable of being widened so as to produce fully separate gene pools.

Sympatric Speciation

Some biologists, while admitting the feasibility of the above

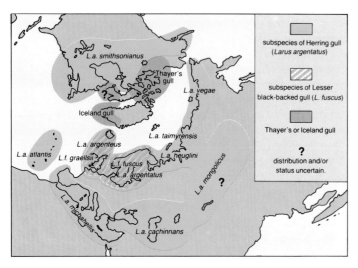

◄▲► **Circumarctic gulls** show a complex pattern of distribution and relatedness that reflects the evolutionary history of an ancestral form separated into a series of refuges during the Pleistocene epoch (2 million–100,000 years ago). There are three distinct species called the Herring gull (*Larus argentatus*) LEFT, Lesser black-backed gull (*L. fuscus*) TOP and Thayer's or the Iceland gull (*L. glaucoides*) RIGHT. These are all morphologically differentiated into a series of races or subspecies, and it has recently been suggested that *L.a. cachinnans* and *L.a. michahellis* may be distinct species too.

Thayer's gull probably evolved in the American Arctic. The pink-legged Herring gull arose in Northeast Siberia, spreading eastwards across America and on into Europe. The yellow-legged Lesser black-backed gull evolved around the Caspian Sea, spreading southwest towards the Mediterranean, and northwest into Britain and Scandinavia. The most distinct forms of these species

(*L.f. fuscus, L.f. graellsii* and *L.a. argentatus* and *L.a. argenteus*) coexist, prevented from interbreeding by ecology and behavior, as are *L. glaucoides* and *L.a. smithsonianus*. This pattern of introgression (ie hybridization) of adjacent races, combined with reproductive isolation at the overlapping extremes is a classic example of a circumpolar-ring species (see map).

schemes, believe that species may be formed where there is no geographical separation of the two diverging populations (sympatric speciation). Somehow a previously freely inter-breeding gene pool must become divided into two essentially non-interbreeding units *in situ.* Many populations display genetic variations, and in some cases these appear to represent adaptations to different facets of the environment. Mating may then occur more often between individuals associating in the same environmental niche than between individuals of different niches. Genes or chromosomal mutants may occur that produce partial sterility when individuals from these two subpopulations try to mate. The hybrids will probably not only be ill-adapted to either of the original environments but will also have lowered reproductive fitness. Accordingly natural selection will tend to reinforce the isolation, slowly increasing the barrier between the populations until they become genetically isolated, even if not geographically isolated.

Advocates of sympatric speciation point to various cases of natural distribution that appear to demand a sympatric explanation. Certain freshwater lakes in Africa, for example, harbor hundreds of closely-related species of cichlid fishes. To derive each of these species by allopatric speciation appears highly implausible. Some recent writers have pointed out, however, that when the population size of an incipient species is very small, it may be difficult to maintain a strong distinction between sympatric and allopatric modes of speciation.

Testing the Theories

Each theory of how species are formed has supporting evidence. But the enormous variety of life forms on earth suggests that the evolutionist must be cautious in attempting to explain speciation by one or two theories alone. There may be many ways in which species are formed, so that one must judge which are the predominant or more important. When viewed in this way, geographical differentiation and isolation seems likely to be the predominant mode.

As manipulating fast-breeding populations in the laboratory becomes easier with modern genetic methods, some speciation theories may be able to be tested experimentally. New advances in molecular biology now permit a very close monitoring of the rate of change in the hereditary DNA itself. Combinations of these methods may open the question of historical changes in gene pools to direct experimental investigation. In the future, science may enable us to speak with greater assurance about the precise ways in which species evolve.   HLC

Macroevolution

Recent techniques (particularly those of molecular biology) have made possible the investigation of the speciation process in a way that was impossible when the only experimental method of study was the attempted hybridization of individuals from different species. However, evolution above the species level (or macroevolution, leading to new genera and higher

taxa) is a study which belongs largely to the domain of paleontology. Although the same processes that produce species differences can also produce greater taxonomic differentiation, there is no general agreement about whether all macroevolutionary changes result in this way or whether other processes sometimes operate. Disputes center around these questions. Are speciation and higher evolutionary changes always adaptive? or is nonadaptive change common?

Major macroevolutionary changes are recognized primarily by study of the fossil record. An example is the evolution within vertebrate animals of adaptations for life on land late in the Devonian period, about 400 million years ago. Emerging from this transition were the amphibians (the group that today includes frogs and salamanders). The fossilized remains of the earliest amphibians reveal a set of bones and curiously complex teeth that are remarkably like those of the lobe-finned fishes that were their obvious ancestors. The branching arrays of bones within the fins became feet. Lungs for breathing air were already present in the ancestral fishes as supplements for their gills, and no special reproductive change was required because amphibians continued to return to the water to lay their eggs.

The fossil record also reveals less dramatic macroevolutionary transitions, such as the origin of the Polar bear (*Ursus maritimus*) from the Grizzly bear (*Ursus arctos*) by speciation.

The Polar bear is unusual among European and American mammals in having virtually no Ice Age fossil record. Only a few scraps of bone, perhaps no more than 20,000 years old, are tentatively assigned to this species. The Grizzly bear, whose fossil remains extend far back into Ice Age deposits, is the only likely direct ancestor of the Polar bear. The speciation event that produced the Polar bear from a population of Grizzly bears was of great potential importance, having opened up a new set of ecological opportunities for the bear family. Whereas the Grizzly bear is an omnivore, with a diet that includes fishes, small rodents, berries and insects, the Polar bear is almost exclusively carnivorous, feeding predominantly on seals, and its teeth are specialized accordingly. Its semiaquatic lifestyle is also probably a new adaptation for the bear family. It might yet give rise to more fully aquatic carnivorous species.

In recent years, the greatest controversy in the field of macroevolution has concerned the tempo and pattern of large-scale evolution. The traditional view has been that major macroevolutionary trends (ones including many species and spanning millions of years) have taken place by the gradual modification of existing species. In recent years, however, many paleontologists have contested the idea that natural selection gradually modifies well-established species in slow but profound ways. The basis for this challenge is the observed stability

of many species in the fossil record. The record reveals that many species of marine bivalve mollusks or foraminiferans (single-celled organisms resembling shelled amoebas) have survived with little change for more than 10 million years. Similar longevities are evident for species of higher land plants. These data and others for groups such as beetles, freshwater fishes and mammals suggest that, barring extinction, an average species may not undergo appreciable evolution (enough to be recognized as a new species) in the course of a hundred thousand or even a million generations.

Those who believe that species stability is the norm tend to reject the notion that large evolutionary changes occurring over millions of years result from the modification of existing species, a process that they regard as normally slow and of relatively minor consequence. They favor instead the idea that evolution is sporadic or "punctuational," being concentrated in spurts that usually occur as new species emerge rapidly from small populations of others. Often, like the Grizzly bear, the parent species continues to live alongside the descendant species. According to this model of evolution, most macroevolutionary trends arise by either or both of two mechanisms. First, successive speciational steps may show a tendency to move in a particular direction. Second, a process of natural selection may occur in which species represent the units. In this sorting process, known as species selection, speciation provides the raw material in the form of diverse new kinds of species.

Certain kinds of species then tend to accumulate in the course of time, because they produce a large proportion of successful new species. They can do this by surviving for an unusually long time, which gives them a high probability of speciating, or by speciating rapidly. The operation of species selection is fully compatible with the occurrence of selection at the individual level. According to the punctuational model, individual selection is most effective in small populations, where by speciation it provides the raw material for the higher-level sorting process of species selection. SMS

▲ **New species for a new land.** The Polar bear ABOVE is thought to have evolved from the Grizzly bear ABOVE RIGHT. This involved the development of features to cope with cold and ice and with a new diet.

◄ **A change of course.** For a century after Charles Darwin published the *Origin of Species* believers in evolution tended to think that it had proceeded gradually, by the accumulation of small changes. It has recently been pointed out that evolutionary rates may vary enormously, so that considerable evolution occurs in sudden bursts at infrequent intervals, so-called "punctuated equilibrium."

Geological time

Gradualism

Punctuated equilibrium

Change in Isolation

Speciation on the Hawaiian Islands

The Hawaiian Islands are a series of volcanic islands in the central Pacific Ocean. The six main islands form a chain extending for 600km (375mi) from northwest to southeast. The oldest of these, Kauai, is 5.6 million years old, and lies at the northwestern end, 120km (75mi) south of the Tropic of Cancer. The volcanoes become successively younger as one proceeds southeast toward the island of Hawaii itself. This island is made up of five volcanoes, and is not only the largest of the group but also the youngest, being less than 400,000 years old. The islands have always been isolated from the nearest continent by at least 3,200km (2,000mi) of open ocean. The volcanic peaks, several of which reach a height of 4,000m (13,000ft) above sea level, form barriers against the warm, moisture-laden trade winds. Thus on the windward sides of the main islands, annual rainfall can be as high as 10,000mm (400in); this is reduced to only a few centimeters on the leeward slopes and beaches. The rain forests are biologically exceedingly complex; the drier ecosystems to leeward are equally interesting. A closer look reveals a wealth of plant species unique to the islands. These in turn harbor a rich abundance of native species of insects, snails and birds. Apart from species introduced by man, both reptiles and land mammals are wholly absent.

Biogeographers and evolutionists have puzzled over this amazing isolated biotic world, and have come to several important conclusions. The first species must have arrived across the ocean after the islands were formed, and therefore extremely recently from an evolutionary point of view. These few individuals must have produced whole populations, which in turn must have evolved quickly and extensively to adapt to the new environments. Some colonists have formed extensive arrays of new species; often these have unique characters, widely different from those of their mainland ancestors. Forms that originally colonized an older island have often undergone a further round of colonization and species formation on the newer islands, thus repeating the original process on an inter-island scale. Especially interesting, therefore, are the species that are unique to Hawaii, the newest of the islands, since speciation must have taken place there within the last 400,000 years; some species are certainly much more recent than that.

Among the animals that abound in Hawaii are *Drosophila* flies, sometimes known as "pomace flies" to distinguish them from recently introduced pests such as the Mediterranean and Oriental fruit flies. *Drosophila melanogaster* is the famous fly of genetics laboratories around the world; like other members of the genus it is economically harmless. It may be found hovering like a tiny gnat over spoiling fruit, attracted there by the growth of yeast. It is perhaps not widely realized that the genus *Drosophila* includes nearly 1,500 separate species around the world. They have adapted to a wide number of different environments, and so are an attractive group for the study of evolutionary processes. What are the kinds of genetic event that accompany the formation of a new species? Such questions are most often asked about some of the more conspicuous mammal and bird species of continents such as Africa. Yet most of these animals reproduce so slowly that it is virtually impossible to study the actual processes of genetic variation in relation to natural selection and population shift—in short, the mechanisms of evolution itself. With *Drosophila* species there are no such problems, since they are generally easy to raise in the laboratory and normally reproduce within six weeks. These flies have therefore been increasingly used for studying the fundamental processes of evolution in populations.

Careful surveys of the insect life of Hawaiian forests have revealed as many as 350 species of *Drosophila* unique (endemic) to Hawaii. These range from small forms to those with a wingspan of up to 22mm (0.9in). Each of the major islands has a unique constellation of species, many of which are found on one island only and so probably evolved there. Researchers have recently concentrated on certain selected species of giant fly that are endemic to the island of Hawaii. Both island and species are very new in evolutionary terms, thus providing an unusual opportunity to observe the first crucial genetic changes that occur in populations as they adapt and evolve into new species. There is extensive genetic variation within individual species, but the genetic changes which are crucial for speciation are those which relate to behavior, and in particular to sexual behavior. Such changes are especially likely to occur in a new population that has been formed from only a few "founder" individuals. The new mode of sexual behavior may then be incompatible with the older one, resulting in a reproductively isolated population—a new species. The successive colonization of new islands in Hawaii has provided many opportunities for such changes to take place, and the unique fauna of each individual island bears this theory out.

Among bird species on Hawaii some of the most interesting are the 28 belonging to the subfamily of Hawaiian finches. They are thought to be derived from a single finch-like species that crossed more than 3,000km (1,860mi) of ocean to colonize the Hawaiian archipelago. From this developed a range of species with specialized feeding behavior and remarkable bills.

HLC

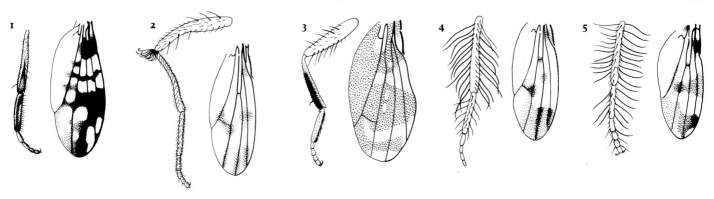

▲► **Examples of species of Hawaiian finches** (subfamily Drepanidinae of family Fringillidae) adapted for different feeding niches. (1) A Grosbeak finch (*Psittirostra kona*), a seed-eater. (2) An akiapolaau (*Hemignathus wilsoni*) which uses its lower mandible to chip wood and the upper mandible to probe for insects. (3) A Kauai akialoa (*H. akialoa*) which probes for insects in thick mosses or deep cracks; bill length about 6.5cm (2.6in). (4) An ula-ai-hawana (*Ciridops anna*), a fruit- and seed-eater. (5) A Maui parrotbill (*Pseudonestor xanthophrys*) which uses its lower bill to chisel into branches. (6) A Crested honeycreeper (*Palmeria dolei*), an insect- and nectar-eater. (7) An iiwi (*Vestiaria coccinea*), a nectar-sipping species with a bill matched to a flower corolla and a tubular tongue for sipping nectar.

◄ **Variety in fruit flies on Hawaii** exemplified in legs and wings. (1) *Drosophila adiastola*. (2) *D. flexipes*. (3) *D. hamifera*. (4) *D. liophallus*. (5) *D. silvarentis*.

Pressure and Resistance

Human activities and evolutionary change

Charles Darwin discussed the remarkable diversity which has been brought about in domesticated animals by artificial selection. He considered the varieties of domesticated pigeon, all derived from the Rock dove (*Columba livia*) and divided them into 11 distinct groups—pouters, carriers, runts, barbs etc. A degree of divergence between the breeds has been brought about in only tens of generations of artificial selection which is equivalent to the differences between species or genera of wild animals subject to natural selection. These varieties are less well known now than they were in Darwin's time, but the same point may be made with respect to other animal breeds.

J. B. S. Haldane proposed a measure of the rate of change in a phylogenetic line, which he called the darwin. This is roughly a change in size of a structure of 0.1 percent per thousand years. While many examples of evolutionary change seen in the fossil record took place at rates to be measured in millidarwins, artificial selection rates would have to be measured in kilodarwins.

For some time after the theory of evolution became accepted it was thought that all natural selection led to evolutionary change but that this was too slow to be experienced at first hand by a human observer. Now, however, it is known that human activities have resulted not only in the conscious artificial selection of new strains of animals and plants, but also in all manner of responses by living creatures to environmental changes that man has brought about. These changes, which truly represent the natural process of evolution, occur at intermediate rates, slower than artificial selection, but fast enough to be observed in hundreds of generations at most.

The classic example is the phenomenon of industrial melanism (darkening). Over 100 species of moths are affected, as well as ladybugs, spiders, bugs and even town pigeons. In each case, genes which make the appearance of the animal darker occur at higher frequencies in areas polluted by smoke and industrial effluents. The European Peppered moth (*Biston betularia*) has a melanic form which is uniform black in color and controlled by a single dominant gene, while the typical form is speckled black and white. The melanic form comprises only 0 to 5 percent of the population in southwest England and West Wales, but over 90 percent in the industrial region of northwest England. Industrial pollution has been present there for 200 years (equivalent to 200 generations in the life of the moth), giving a maximum time for this divergence to have occurred. In Leicestershire, high frequencies of melanic individuals of a wolf spider, *Arctosa perita*, have been found on coal spoil tips. They belong to a coastal species, and must be derived from nonmelanic populations introduced with coastal sand not more than 100 years previously. In South Wales there is a very high frequency of melanics in several different types of insects living near a pollution source in the form of a "smokeless fuel" plant which commenced operation only 45 years ago.

Rapid change in gene frequency shows that the new environment has imposed strong selection in the species involved. The change in frequency of the dark form of the Peppered moth shows that in a polluted area a melanic moth has a 50 percent greater chance of survival than a typical. Undoubtedly, one of the principal causes of selection in this case is predation by birds, which find the typicals easier to see than the melanics on the blackened and lichen-free bark of trees in industrial areas. In other cases, the causes are more obscure, but differences in metabolic rate may be involved.

With the introduction of DDT and penicillin in the 1940s mankind again embarked on a major program to modify and control the environment. Many other pesticides and antibiotics have since been synthesized, substances previously unknown in the natural world. Until the harmful consequences were recognized, both pesticides and antibiotics were used in a profligate manner. Very soon, however, it became apparent that the insect pests and disease carriers, and the disease organisms themselves, could undergo genetic adaptation to combat the new control agents.

One striking example has been the development of resistance by the House fly (*Musca domestica*) on farms in Denmark, where intensive chemical control has been practised. DDT was first used in 1945, to be followed by a succession of other insecticides as resistance developed. At present the flies exhibit some degree of resistance to all known controlling agents.

Such examples, where man has performed unwitting selection experiments, demonstrate the facility with which organisms can adapt. Ultimately the process depends on the selection of mutants produced at random during cell replication. Very rapid change may sometimes result from recombination of genes already present. Recently, when studying drug resistance in bacteria, scientists have discovered previously unknown mechanisms by which genes are transferred from one individual to another, increasing further the rate of adaptive response. The slow changes exemplified by the fossil record therefore result from the relative constancy of the environment and not from any limitations imposed by genetic structure or mutation rate.   LMC

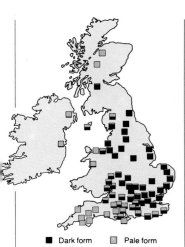

■ Dark form   □ Pale form

▲ **The dark form** of the Peppered
moth: its frequency in populations
in Britain, 1952–56. The dark form
was prevalent in industrial areas.
With greater control of air pollution
the area of high frequency has since
contracted towards the northeast.

▶ **Conspicuous and cryptic:** typical
and dark forms of the Peppered
moth on a soot-stained tree, such as
are found in polluted urban areas.

▼ **Pigeon breeds** exhibited in 1864
by a London pigeon club of which
Darwin was a member. Pigeon
breeds, he suggested, showed the
variety that could arise from a
single species.

▶ **Genetic resistance.** Inherent
genetic variability has enabled
insects to resist control by
pesticides. In Denmark the House fly
became capable of resisting all
known controlling agents. In some
cases resistance developed in the
extraordinarily short time of one
year, whilst resistance to other
substances took 10–20 years to
build up. The resistance may be
achieved by blocking entry of the
poison, by excreting it as fast as it
enters, or by breaking it down into
harmless compounds within the
body. Many different genes may
therefore be involved in coping with
the same insecticides.

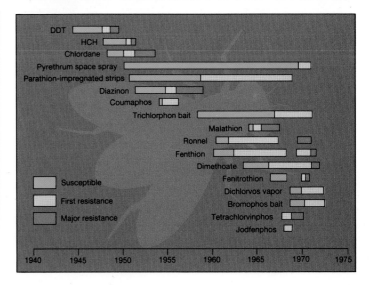

Man
and Evolution

When Charles Darwin's theory of evolution was published one feature seemed to make it particularly offensive: the idea that humans had emerged from the natural world. Since then the findings of archaeologists have outlined our recent evolutionary history, though our exact relationship to the other primates remains uncertain and disputed. The investigation of variability within mankind has permitted a new assessment, based on biochemistry and genetics rather than on visually observable features, of the extent of differences between human races.

Controversy was also generated by many other aspects of Darwin's theory. At first a major weakness was his inability to account for continuity and change within living organisms (as opposed to the elimination of disadvantageous ones). The rediscovery of Mendel's work enabled one of Darwin's major problems to be dealt with, in that variation was not eroded by blending in each generation. However, this fired further controversies about the extent to which genetic mutations drive evolutionary change. The most persistent opponents of evolution have been "creationist" Christians, but there is less unanimity in their criticisms than at first appears.

◀ **The dominant animal species** on earth: *Homo sapiens.*

MAN

Man's division into races according to physical differences. . . Blood groups vary independently of geography or race. . . Piecing together man's ancestry from fossil evidence. . . From Australopithecus *to* Homo sapiens. . . *Evidence of natural selection in man's development. . . Man's greatest adaptation is adaptability itself*

M AN is certainly an animal, but not just another animal. By almost any criterion he must be judged the most successful animal of all time, who is able to determine not only his own evolution but also that of practically every other organism on earth.

Yet for much of his evolutionary history man was a rare animal. Even when he was indistinguishable in skeleton—and probably in most other structures, including the brain—from present-day man, his total numbers do not appear to have exceeded a few thousand at any one time. In this period, between 40,000 and 10,000 years ago, he was living the life of a "hunter-gatherer," rather like that of the present-day Bushmen of the Kalahari Desert or the Australian aborigines in Arnhem Land. The first great change came with the Neolithic Period, when humans first began to practise the cultivation of plants and the domestication of animals. From then on the numbers of people have increased almost exponentially; the species has spread over the whole land surface of the globe; and population densities can exceed 800 per sq km (2,000 per sq mi) even in agricultural areas, rising to over 40,000 per sq km (100,000 per sq mi) in certain urban areas.

While ability to create an increasing food supply has certainly been the basic determinant of population growth, this is only part of that very special attribute of man, namely the possession of culture, including language and the ability to create artifacts. Although rudimentary elements of culture have been identified in other species, especially in man's closest relatives the primates, even the simplest of human societies possesses social dimensions many times greater than anything that is observable in any other species.

The Races of Man

All present-day human beings belong to a single species, *Homo sapiens*; by this is implied that they could all mate with one another and produce healthy, fertile offspring. Since the dramatic increase in travel in recent decades, most of the world's populations have been brought into contact with one another, and have thus clearly demonstrated the validity of the one-species concept. Nevertheless there is as measured biochemically an enormous amount of genetic variation within the species, and it is probable that, identical twins apart, no two individuals who have ever lived have had the same genetic constitution.

The variation that first attracted the attention of anthropologists was variation in characters that are visually obvious, such as skin color, eye color, hair form and body dimensions, especially stature and head shape. Many of these characters show quite striking geographical variation, and they tend to

▲ ▶ The racial groups of mankind. The map shows the distribution of aboriginal populations in the late 15th century with approximate locations of main subgroups. Around the map in clockwise order are representative members: Mongoloid (Cofan Indian from Colombia); Caucasoid; Negroid (from Cameroun); Mongoloid (from Laos); Australoid (from Australia); Khoisan.

Racial Groups

- Australoid
- Negroid
- Khoisan
- Caucasoid
- Mongoloid

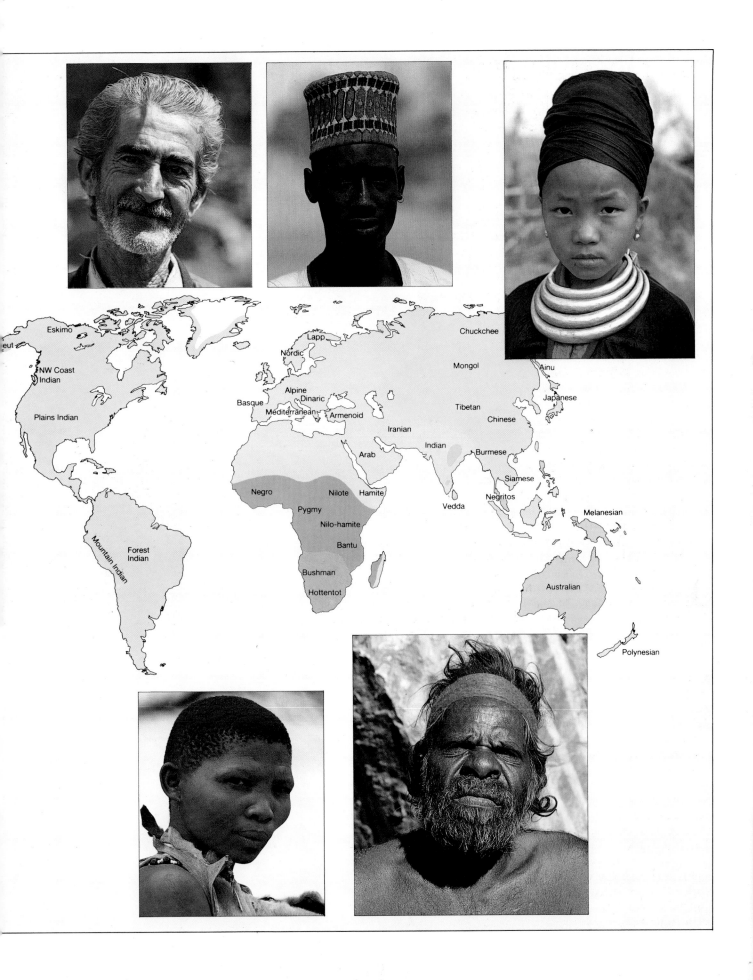

Eskimo
eut
NW Coast
Indian
Plains Indian
Mountain Indian
Forest
Indian
Lapp
Nordic
Basque
Alpine
Dinaric
Mediterranean
Armenoid
Iranian
Arab
Negro
Pygmy
Nilote
Hamite
Nilo-hamite
Bantu
Bushman
Hottentot
Chuckchee
Mongol
Ainu
Tibetan
Japanese
Chinese
Indian
Burmese
Siamese
Negritos
Vedda
Melanesian
Australian
Polynesian

differentiate human populations on a continental basis. Moreover, the distribution patterns of many of the characters are similar or concordant, so that there appear to be major geographical groups of people with a number of characters in common. These observations formed the basis of the various classifications that were made of human populations into races. The peoples of Europe and western Asia tend to be fair-skinned with wavy hair and were identified as Caucasoid; those of Africa south of the Sahara tend to be dark-skinned with curly hair and were called Negroid; and those of East Asia and the Americas are yellow-skinned with straight hair and constitute the so-called Mongoloid races. Other major categories that were commonly recognized were the indigenous peoples of Australia—the Australoids—and the Bushmen and Hottentots of southwestern Africa—the Capoids or Khoisans. Then various finer classifications were made of the major groups into sub-groups, such as Nordics, Alpines and Mediterraneans (types of Caucasoid). Even then many populations were hard to fit easily into the groups that had been recognized. The Ethiopians with their Caucasoid features and Negroid pigmentation, the Melanesians, the Vedda populations of South India, and the Ainu of Japan, are examples of such troublesome groups.

Matters became more difficult with the discovery of a whole host of biochemical variations. Most of these have been found in blood because it is easy to obtain and analyze, but other tissues are likely to show similar variations. The first of the variations in blood are the various blood-group systems: the 11 red-cell systems, including ABO and Rhesus, and the white-cell systems such as the transplantation antigens HLA. There are also variations in many of the proteins that are to be found in human sera: in the haptoglobins, which combine with the heme of hemoglobin and facilitate its conversion into bile salts; in the transferrins, which transport iron; in the immunoglobulins; and in many others. Lastly, there are variations in a whole host of enzymes that typically occur in blood

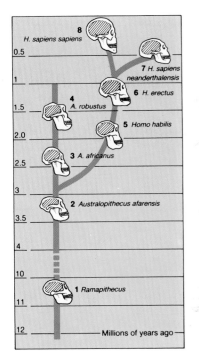

▲ **Chronology of human evolution.** See following captions for key.

◄► **The evolution of man.** (1) *Ramapithecus.* (2) *Australopithecus afarensis.*
(3) *A. africanus.* (4) *A. robustus.* CONTINUED ►

cells: phosphoglucomutase, adenylate kinase, and adenosine deaminase. Molecular biology is also now revealing endless variety within the DNA of cell nuclei and mitochondria.

These variations are in a number of ways easier to understand and more revealing than the visually obvious characters. In the first place, the expression of each character variation is due solely to genetic causes, and is not affected by the environment in which individuals develop. Thus the variation remains unaffected by "nurture," unlike stature, which is affected by nutrition and disease, or skin color, which tends to darken when exposed to ultraviolet radiation. Secondly, the differences between individuals are discrete (eg either A, B, O or AB in the ABO blood-group system), and not continuous as they are in most morphological and physiological characters (eg stature may be anything between short and tall). This is because the mode of inheritance is relatively simple, and being simple has been fully worked out. One can therefore characterize populations in terms of the frequencies of different genetic variants within them. Thus the frequency of the O blood-group gene in England is around 65 percent. Such an analysis is not possible for any of the genes determining continuous traits. Indeed, individuals of the same stature may well have different genes.

When biochemical and serological variation is examined in human populations, a number of patterns appear. Most importantly, the amount of the variation is colossal. It is estimated that human beings are heterozygous (ie possess two different variants) for 15 to 20 percent of their genes, and something of the order of 25 percent of genes are polymorphic (ie have common alternative alleles). This is why everybody must be

5

6

genetically unique, even if the total number of genes is as low as 50,000. With 14 red-cell enzyme systems alone, the likelihood in Britain of randomly discovering two identical genotypes is 1 in 32,000. Most of this genetic variation, however, lies within each and every human population, and the variation between populations, ie the racial variation, is comparatively small, of the order of 10 to 15 percent. Populations world-wide differ mainly in the frequency with which the same genes occur, rather than in having totally different genes. In other words, in biochemical characters Europeans differ only slightly more from Africans or Asians than they do from their immediate neighbors. What is more, when variations are plotted geographically, it is found that populations with very different gene frequencies are usually connected by populations showing a gradation of intermediate gene frequencies. The resulting gradients of gene frequency are known as clines. Moreover, different systems are rarely concordant in their geographical distribution, often showing totally different patterns. All this has led to the general discrediting of the race concept in man, and it has been said that "there are no races, only clines." However, common experience suggests that this is going too far: racial variation exists, even if classifying people into races is an arbitrary exercise.

Fossil Man

The main reason for similarities in gene frequencies between populations is recent common ancestry; populations which are markedly different from one another have usually separated a long time ago. Tracing the ancestries of living human groups, generally using levels of genetic similarity as a major source of information, has long been a preoccupation of anthropologists. But they are also concerned with determining the origin of the species itself, and with tracing the complete phylogenetic history of hominids—the particular family of primates to which *Homo sapiens* belongs. The evidence for this comes essentially from the fossil record.

It is not known with any certainty when the hominid lineage separated from the line that gave rise to the present-day great apes, or pongids. Until a few years ago, it was thought that the division occurred in the early Miocene from a primitive

◄► **The evolution of man**
continued. (5) *Homo habilis.*
(6) *H. erectus.* (7) *H. sapiens neanderthalensis,* the Neanderthals.
(8) *H. sapiens sapiens.* One notable feature of human evolution was an increase in body size. *Ramapithecus* and the australopithecines were about 1.1m (3.5ft) tall. The Neanderthals were shorter than most modern human races.

7

8

dryopithecine ape. Evidence from molecular biology, however, suggests very close similarity between man and the two African apes: the gorilla and the chimpanzee. This occurs in many serum proteins, hemoglobin and even DNA and has been interpreted to mean that common ancestry is as recent as 4 to 5 million years ago. There is now some paleontological support for this view.

One possible early hominid ancestor was *Ramapithecus*—first discovered in India, but now also known to have existed in East Africa and the Near East. In jaw and tooth characters (from which it is mainly known) it shows some human characters such as small canine teeth and parabolically-shaped tooth rows. Recently, however, a connection between *Ramapithecus* and the orang-utan has been proposed. Here it needs to be noted that *Ramapithecus* varies in both time and place, and the various fossils attributed to it tend now to be referred to a wider taxon—the ramapithecines.

Apart from a few bone fragments found in 5–4 million-year-old deposits, the earliest definite hominid fossils are dated to just over 3 million years ago. They come from Hadar in Ethiopia and Laetoli in Tanzania, and have been referred to *Australopithecus afarensis*. They were small creatures, little bigger than a medium-sized monkey, but showed the skeletal development that is associated with erect posture and bipedal gait—man's most distinctive anatomical characteristics.

From somewhat later deposits in East and South Africa came a whole host of fossils which are also attributed to *Australopithecus*. Some of the more productive sites are Koobi Fora on Lake Turkana (formerly Lake Rudolf) in Kenya, Olduvai Gorge in Tanzania, and Sterkfontein and Swartkrans in the Transvaal. The first australopithecine was discovered in 1924 at Taung in Botswana. There is considerable diversity within the australopithecines. Not surprisingly, early forms are more primitive than later ones. There is also a clear distinction between a heavy, robust type (*A. robustus*), which was probably vegetarian in diet, and a more gracile form (*A. africanus*), which may well have been at least partly carnivorous. The essential feature of all australopithecines however is that they combined bipedal gait with a relatively small braincase and relatively large jaws and teeth. Despite these ape-like characteristics of the skull, which merely represent the retention of ancestral traits, there are a number of undoubtedly human elements in the morphology.

Depending somewhat on definition, australopithecines survived to nearly 1 million years ago and were essentially African. They almost certainly gave rise to the later hominids, all of which are nowadays attributed to the genus *Homo*. Some intermediate forms have been found in East Africa, which have been variously called *Homo habilis*, *Australopithecus habilis* or *A. africanus habilis*! The essential features distinguishing *Homo* from *Australopithecus* are proportionately larger brain size compared with the (increasing) body size, and reduction in the jaws and teeth. The ultimate dividing line is, of course, arbitrary; indeed if the fossil record were complete such a line would have to separate parent and child!

Some of the earliest representatives of *Homo* were found in the Far East, particularly Java, and were first attributed to *Pithecanthropus*. Java man had a cranial capacity of around 1,000ml (60cu in), almost intermediate between *Australopithecus* and *Homo sapiens*. Very similar to Java man but rather later and more refined are the remains of Peking man from Choukoutien in China. These Asiatic hominids are today referred to *Homo erectus*. Forms broadly similar have been discovered in Europe and Africa and some of the East African forms may be earlier than the Javanese remains. They are also usually referred to *Homo erectus*, but it has recently been suggested that they may represent the actual populations which gave rise to *Homo sapiens*.

Forms which are attributed to our own species first appear in the fossil record about 500,000 years ago. These early *H. sapiens* tend to have larger brains than *H. erectus*, but they still have thick skull bones, poorly-developed foreheads and typically a heavy bony bar above the orbits for the eyes. Most of the earliest examples are known from Europe, including the Swanscombe remains from Kent, England, but the forms were probably quite widely distributed in the Old World. One fairly late group occupied Europe and surrounding areas at the time of the beginning of the last glaciations; it comprises the so-called classical Neanderthals, in which the thick skull bones, poorly-developed forehead and heavy bony bar above the eyes seem particularly pronounced. This is something of a puzzle, and various explanations have been offered, including the idea that classical Neanderthals represent a rather slowly evolving lineage in northern Eurasia which moved into Europe during the Ice Ages. What is certainly known is that classical Neanderthal populations were replaced in Europe about 50,000 years ago by forms skeletally indistinguishable from ourselves. Around the same period, modern forms of man were appearing throughout the Old World including Australia. They first entered the Americas across the Bering land bridge considerably later, about 25,000 years ago.

Adaptation and Natural Selection in Man
Perhaps the most significant step in human evolution was

▲▶ **Human adaptability.** Humans are physiologically highly adaptable—able to acclimatize to heat, cold and high altitude, to develop immunity to diseases, and to satisfy nutritional needs from an endless variety of foods. Growth and development are also adaptively modifiable and produce characteristics appropriate to the growth environment. ABOVE bushmen in southern Africa. RIGHT an eskimo. Each has a physique appropriate for the environment.

when we stopped living in trees, the habitat generally occupied by primates, and adapted ourselves to living on the ground by walking on two feet. This form of locomotion has no general advantages in itself, but it completely frees the forelimbs to do other things. Being of primate origin, early hominids had already adapted their forelimbs as grasping organs, which, when freed from their role as organs of support and movement, could easily be used for carrying things, for using tools and later for making them as well. Hand and brain could then evolve together.

There can be little doubt that natural selection has been the driving force behind the changes documented in the human fossil record. Bipedal gait, brain size, and the capacity to manufacture tools and learn language, all clearly had survival value and improved Darwinian fitness. But what of the enormous genetic variety which is found in all present-day populations? Is this also signficant from the point of view of natural selection? This is a matter of some central concern and controversy in population genetics today. The variation in the biochemistry of human blood seems to have no functional significance in present-day environments, so that the alleles which determine, for example, whether an individual has blood group O or A are nowadays of neutral survival value.

However, despite the considerable difficulty of detecting selection in human populations, some examples have been found of where natural selection appears to operate on human genes. One of the best examples is the high incidence of sickle-cell hemoglobin in Africa, around the Mediterranean, and in India and its environs which is related to resistance to the malaria caused by *Plasmodium falciparum*. There are several

other genetically controlled variants which also appear to provide resistance to certain types of malaria. These include hemoglobin C in West Africa, hemoglobin E in Thailand, hereditary persistence of fetal hemoglobin in Africa, and the β-thalassemias with a wide distribution in the tropical world. One variant gene affecting enzyme levels, causing glucose-6-phosphate dehydrogenase deficiency, also appears to afford malarial protection. Malaria may well be a relatively recent human disease in evolutionary terms, and may therefore provide some insights into how the human genetic constitution evolved. All these protective mechanisms carry some sort of disadvantage—though not equally so—and eventually some better systems may arise. That is, of course, provided that malaria is not first eradicated, or that it does not evolve so as to counteract these protective mechanisms—in which event scientists would be left wondering about the significance of all these hemoglobin variants!

Quite a number of diseases have been found to be associated with various types of polymorphism, particularly the ABO blood groups. But most of these do not appear to be acting with sufficient intensity to have any selective influence; many only arise after the normal age of reproduction, and some act in the wrong direction.

However, there are other examples of genetic selection to be found in human populations. A particularly interesting one concerns the persistence into adulthood of the enzyme lactase in the lining of the intestine. In most mammals this enzyme, which breaks down the milk sugar lactose, is only produced during the period that the infant is being fed on its mother's milk. The same is true in many human groups; but in those populations which rely heavily on milk and milk products for their nutrition, such as the Europeans and the cattle herders of Africa, the enzyme persists into adulthood. This appears to be controlled by a simple gene, and represents an excellent example of a cultural factor—namely animal domestication—acting as a selective agent.

Many of the complex character variations which were first studied by anthropologists are also clearly related to selective agents in the environment. Dark pigmentation protects the skin from the damaging effects of ultraviolet radiation; while in the cloudy and sunless environs of northern Europe it is advantageous to permit as much of the limited UV as possible to penetrate the skin, since this helps the body to produce more vitamin D. (Shortage of this vitamin causes rickets in children and osteomalacia in adults.) Moreover, much of the geographical variation in body size and shape may be related to thermal adaptation.

Notwithstanding these genetically-determined adaptive differences between human populations, all human beings have a remarkable ability to adapt to the environment in which they are living in order to survive. This is known as adaptability, and high adaptability is a species character of *Homo sapiens*. This means not only that human beings are found over the whole land surface of the earth, but also that any one person can live practically anywhere. But man's greatest adaptability lies in his behavior, which shows an infinite ability to respond to a whole complex of different environments, both in the natural world and in social terms. It is surely this behavioral adaptability, more than anything else, which has been responsible for man's evolutionary success.   GAH

CONTROVERSIES OF EVOLUTION

*Mendelists and biometricians. . . Biometricians suggest
that evolution is continuous. . . Geneticists see inherited
variation as discontinuous. . . Neo-Darwinism:
paleontologists and geneticists. . . Some see evolution
functioning independently of natural selection. . . Later
evidence supports the Darwinian view. . . Philosophical
and theoretical objections to natural selection. . .
Cladism. . . Neo-Lamarckism and creationism hark back
to before Darwin. . . Evolution, ethics and Christianity. . .
Is altruism discouraged by natural selection?. . .
Ultimately there is no conflict between Christianity and
evolution*

EVOLUTIONARY ideas often produce controversy, which
sometimes involves legitimate scientific debate, but frequently merely ignorance or preconception. Notwithstanding
it was not debate but straightforward geological discoveries in
the 18th and early 19th centuries that prepared the way for
Darwin's work. By the time the *Origin of Species* was published
(1859) it was clear that many changes had taken place in
animals and plants since they first appeared on earth: knowledge of the distribution of organisms had reached the point
where it was clear that many species had become extinct; the
main outline of the fossil record was well-established; and cal-
culations of rates of erosion and deposition showed that the
earth seemed to have been in existence very much longer than
indicated by Archbishop Ussher's date for creation of 4004 BC
(based on adding up the ages of the Old Testament patriarchs).
In retrospect Darwin's easily understood and testable theory
for the mechanism of evolutionary change was his most
important contribution, but at the time he convinced most
people of his ideas because his explanation of the causes of biological variety, and in particular the reasons why species were
not uniformly distributed over the whole world.

Charles Darwin himself was well aware of many of the problems with his ideas; he had recognized the role of natural selection in producing evolutionary change in the late 1830s after
his return from the *Beagle* voyage, and thus he had nearly 20
years to think about them and discuss them with his friends
and correspondents before the *Origin of Species* was published.
In chapters six and seven of the book he dealt with "difficulties"
and "miscellaneous objections." His main points concerned the
nature of species and the effectiveness of natural selection; in
a later chapter he discussed gaps in the fossil record. He recognized that ignorance about the ways variation is maintained
was a key weakness in his theory, and the causes of variation
are repeatedly referred to in his book. This problem was in principle solved by the embryologist August Weismann, who pointed out (1883) that, in higher animals at least, the reproductive
cells are distinct from nonreproductive ones and thus largely
unaffected by environmental influences. The work of Gregor
Mendel (1866) laid the basis for our modern understanding
of the mechanism of inheritance and the causes of variation.

Mendelists and Biometricians
Darwin accepted the assumption of his time that the characteristics of both parents "blend" in their offspring. This meant that
if a new variant arose it would have only half its expression
in the offspring, one-quarter in the grandchildren (because it
would almost certainly cross with the nonvariant form), and
so on. New variation would have to arise at a high rate if it
was to persist long enough to affect natural selection. In the
Variation of Animals and Plants under Domestication (1868),
Darwin put forward his "provisional hypothesis of pangenesis"
in an attempt to account for this.

Darwin's cousin, Francis Galton, carried out extensive
investigations into human heredity. He was impressed by the
ways in which size, disease, temperament, and so on were
transmitted in families. In *Natural Inheritance* (1889) he set out
a "law of ancestral heredity," developing statistical techniques
of correlation and regression, which remain central to comparing individuals and groups. These concepts were seized upon
by a mathematician, Karl Pearson, and a zoologist, W. F. R.
Weldon. Weldon wrote, "The questions raised by the
Darwinian hypothesis are purely statistical, and the statistical
method is the only one at present obvious by which that hypothesis can be checked"; he collaborated with Pearson in seek-
ing methods to test and extend Galton's law. Together, they
developed the subject of biometry (ie the measurement of
organisms).

Then in 1900 Mendel's work was rediscovered and quickly
confirmed to apply to a wide range of organisms, including
man. It had two implications for evolution. First, inherited factors do *not* blend, but persist unchanged through the generations, so that the persistence of variation was no longer a
problem. Second, Mendel's experiments appeared to show that
inherited variation was discontinuous, whereas the biometricians regarded continuous variation as the raw material of
evolution. The years between 1900 and 1906 (when Weldon
died from pneumonia at the age of 46) witnessed increasingly
bitter confrontations between the biometricians on the one
hand and the Mendelists, led by the Cambridge geneticist William Bateson, on the other. However, the work of T. H. Morgan
and his colleagues on sex determination and chromosomal
linkage and recombination in *Drosophila melanogaster* showed
that the inherited factors are carried on the chromosomes, and
that the events of cell division form the physical basis of
heredity. This contributed to the ascendancy of the
Mendelists—or, as we would call them today, the geneticists.
Notwithstanding, the split between the supporters of continuous and discontinuous evolution persisted, and was only
resolved in the 1930s.

Neo-Darwinism: Paleontologists and Geneticists
In the early 1900s the importance of natural selection in producing evolutionary change was thought to be slight; it was
the origin of variation rather than its maintenance that was
stressed. Thomas Huxley and William Bateson both believed
that continuous variation produced too little effect to generate
significant selection pressures, and Galton believed that the
selection for such variation soon reached a limit because of the
counteracting effect of regression. Wilhelm Johannsen, a
Danish botanist, showed in his experiments on beans in 1908
that selection for weight (of the beans) within pure (ie self-

▲ **Man reunited with his long-lost relative,** a cartoon reaction to Darwin's theory about the descent of man.

fertilized) lines of descent had no effect, and argued from this that continuous variation was not inherited and was therefore of little importance in evolution. H. de Vries (one of Mendel's rediscoverers), working on the Evening primrose, *Oenothera lamarckiana*, concluded that new species were formed by distinct mutations; his views were particularly influential at the time.

We now know that de Vries's *Oenothera* mutations were mostly due to chromosomal rearrangements, and were not mutations in the modern sense. However, a generation of biologists grew up convinced by de Vries's beliefs that evolution was "driven" by mutations, with natural selection taking a minor role. This happened at a time when paleontologists were building up an increasingly confident picture of evolutionary changes in fossil strata. It seemed clear to them that much change was continuous and progressive; evolutionary "jumps" were not found when the record was continuous over long periods. The mutations being studied by laboratory geneticists appeared to have nothing in common with real evolution. Darwin's own emphasis on gradual evolution was continued by the paleontologists, following the line of the biometricians.

During the 1920s the gap between paleontologists and geneticists widened. As knowledge of mutations (in today's sense) grew, it seemed that they almost invariably produced deleterious traits which were inherited as recessives; that is, the trait was expressed only when the allele of the gene was

present on the chromosome from both parents, whereas adaptively useful traits were virtually always dominantly inherited. This was a period which led to widespread disenchantment with classical Darwinism, and the propounding of a variety of other theories of evolutionary mechanism: Berg's "nomogenesis," Willis's "age and area," Smut's "holism," Driesch's "entelechy," and others. None of these were satisfactory, and all depended for change on a somewhat mystical inner urge (or *élan vital*). It is unfortunate that three standard histories of biology, by Nordenskjold, Radl and Singer, were written during this time, and perpetuated the idea that evolutionary theory is an illogical mess.

At the height of the Mendelist–biometrician controversy, a mathematician, Udny Yule, showed that if one no longer assumed that dominance was always complete, and if the effect of environment on variation was taken into account, the conflict between Mendel's findings and Galton's law disappeared. The importance of the environment became increasingly clear from plant-breeding experiments by E. M. East and others, and from the demonstration by embryologists (E. S. Goodrich in *Living Organisms*, 1924, Julian Huxley, Gavin de Beer, etc) that the primary effect of a gene may be considerably modified in expression during development.

The question of dominance was taken up by R. A. Fisher. Using Pearson's measurements on man, Fisher dissected the factors involved in characteristics affected by many genes, examining statistically the consequences of gene interaction, nonrandom mating, and linkage, on the observed correlation between relatives. By comparing the correlations between parents, he was able to disentangle the contributions of dominance and environment to the total variation. This led Fisher to believe that dominance was a result of interaction between genes rather than an intrinsic property of the gene itself. He showed that beneficial characters will be selected for dominance, and will spread by natural selection to replace a previous trait. Conversely, alleles deleterious in the heterozygote (ie when inherited from one parent only) are likely to be transmitted only in combinations where their effect is least; that is, there will be selection for a small heterozygous effect, which means in the direction of recessivity and thus minimal expression in the individual.

Fisher put forward these ideas in 1928 from first principles. They were received skeptically because of the difficulty of believing that selection pressures could be strong enough to allow the spread of genes which did nothing else but modify dominance. Before the 1950s natural selection was generally assumed to be a weak force producing effects of 1 percent or less (ie 100 individuals with an advantageous trait would survive or breed, compared with 99 individuals not possessing the trait). If this was so, second order selection depending on interactions between genes (and hence including those genes whose main effect is to modify the expression of other genes, such as in dominance modification) would have such a small influence that it could be ignored. However, Fisher himself was able to demonstrate the genetic modification of dominance, by crossing domestic poultry with wild jungle fowl for five generations. This changed the mode of several characters. Fisher reasoned that the dominance of the traits he studied had been attained during domestication and that this could be changed as the genetic effects of domestication were removed by outcrossing.

Fisher's theory has been proved right in a number of experiments with both plants and animals. Although it may not apply for every trait, it has had considerable significance in bridging the gulf between paleontologists and geneticists. In effect, the theory provided the genetic basis for the understanding between disciplines that was needed. Fisher's work, together with that of other theoreticians such as Sewall Wright and J. B. S. Haldane, was a major element in the reevaluation of evolutionary ideas that has come to be called the "neo-Darwinian synthesis."

As late as 1932 T. H. Morgan asserted that "natural selection does not play the role of a creative principle in evolution," but 10 years later almost all biologists were agreed on an evolutionary theory based firmly on Darwin's own ideas combined with subsequent developments in genetics. This coming together was described by Julian Huxley as the Modern Synthesis in a book of that name published in 1942. The synthesis was in fact first set forth in books by three Englishmen, R. A. Fisher's *The Genetical Theory of Natural Selection* (1930), E. B. Ford's *Mendelism and Evolution* (1931), and J. B. S. Haldane's *The Causes of Evolution* (1932). It was consolidated in three works from America, T. Dobzhansky's *Genetics and the Origin of Species* (1937), Ernst Mayr's *Systematics and the Origin of Species* (1942), and G. G. Simpson's *Tempo and Mode in Evolution* (1944). As Mayr wrote, the synthesis came not from one side being proved right and the other wrong, but from "an exchange of the most viable components of the previously competing research traditions." The dissensions and difficulties of the 1920s and 1930s can, with the benefit of hindsight, be seen as a result of increasing specialization by biologists. Darwin himself collected as many facts as he could from the whole of biology, but by the end of the 19th century each of the disciplines within biology had grown so much that the glue binding Darwin's original synthesis had begun to come apart. The significant movers in the neo-Darwinian synthesis were scientists who broke out of their own specializations and transferred ideas from one discipline to another.

Non-Darwinian Evolution: Neutralism and Selection
There is a recurring argument that evolution cannot depend upon mutation, because mutations are "always harmful." This is a false argument. Although most mutations are harmful to their possessors, since they are random changes in a functioning organism, some are undoubtedly beneficial. There are now many examples of this in natural populations. In addition the value of many new mutations has been shown experimentally in the large amount of work that has been done on the consequences of mutation induced by radiation and other agents.

Mutations provide the raw material for evolutionary change, and it is necessary that their effects on populations should be fully examined. This was attempted during the 1950s and 1960s when the achievements of the neo-Darwinian synthesis were formalized in theoretical arguments, in particular by H. J. Muller (one of T. H. Morgan's colleagues in *Drosophila* research) and J. B. S. Haldane. Muller reasoned that there is an upper limit to the amount of variation that may be carried by an individual or population. We all have a burden of deleterious alleles produced by recurring mutation; these form a genetic "load," bearing us down. Clearly there is a limit to the number of such genes which any individual can carry and still survive to reproduce. In so far as these genes reduce health and fertility, they are eliminated by natural selection.

In 1957 J. B. S. Haldane called attention to the "cost" of natural selection. This occurs whenever an allele spreads in a population. Those individuals possessing the allele that is being replaced will be less fit than those possessing the allele that is spreading. A population can only tolerate a limited number of substitutions going on at the same time; otherwise too high a proportion of individuals will die, and a viable population cannot be maintained. Haldane's "cost" represents a "substitutional load" in Muller's terminology; Muller's own emphasis should be called a "mutational load." There are other elements to the total load; they all contribute to the idea that there is an upper limit to levels of genetic variation.

In the mid-1960s the application of electrophoresis to a series of proteins studied in a large number of individuals from interbreeding populations showed that very high levels of inherited variation occur in nature: although a few species have a low level of variation, most have a large amount of heterozygosity, ie many gene-loci at which different alleles are inherited from the two parents, ranging from 3 or 4 percent in large animals to around 25 percent in small ones. Genetic variation is far more common than was previously assumed; the theoretical basis of neo-Darwinism, which assumed that variant genes impose a load on their possessor, seemed to be destroyed.

The simplest way out of the variation dilemma was to assume that the protein variants detected by electrophoresis were selectively neutral; that is, that they had no effect on the breeding success or survival of their carriers. The main argument for this conclusion, known as "neutralism," was the alleged accuracy of the "protein clock."

The protein clock is based on the assumption that the substitution of an amino acid at any place in a protein chain is likely to take place at random, and therefore the number of amino acid differences between two species can be regarded as a measure of the time since the species shared a common ancestor. A genealogical tree constructed on this premise would show quantitative degrees of relatedness. When this was done, there appeared initially to be remarkable uniformity in the rates of substitution in completely different evolutionary lines.

However, it soon became clear that different proteins have rates of change differing by nearly a hundredfold. Biochemists speculated about the functional importance of different parts of protein molecules, suggesting that there were "functional constraints" to some parts, whilst other parts were apparently dispensable. This could mean that changes at "unimportant" sites within a molecule might proceed more rapidly than at others, but that other changes would be slowed down. These ideas have been valuable in probing the functional morphology of some well-studied proteins (notably the cytochromes, globins and immunoglobins), but destroy any idea of an accurate protein clock.

Evolution, Ethics and Christianity

There have been many attempts to show that ethics or moral standards have been formed by the evolutionary process itself, but all such attempts have been uniformly unsuccessful. The difficulty is that natural selection means that individual survival and reproduction is rewarded, whereas altruism is punished.

Darwin himself recognized this, commenting: "It is extremely doubtful whether the offspring of the more sympathetic and benevolent parents, or of those who were the most faithful to their comrades, would be reared in greater numbers than the children of selfish and treacherous parents belonging to the same tribe. He who was ready to sacrifice his life, as many a savage has been, rather than to betray his comrades, would often leave no offspring to inherit his noble nature."

The genetical disadvantage of altruism is not absolute, because an altruistic non-breeder may be able to help a relative to raise more offspring than he or she would have done if not helped. This *kin selection* may compensate for the evolutionary disadvantages of an individual's altruism (or unselfishness), and allow inherited traits to spread which favor group rather than individual survival. The study of such mechanisms is called sociobiology.

Can kin selection explain the so-called "higher qualities" in man—love, joy, patience, kindness, etc? Julian Huxley believed that "the human species has emerged from its biological phase of evolution, and entered into a psycho-social phase"; more recent scholars (notably E. O. Wilson and Peter Singer) have regarded genetical processes as wholly sufficient to explain moral and spiritual characteristics. However, no moral trait has been shown to be inherited.

Attempts to explain human nature by natural processes have traditionally been resisted by most religions. Disraeli wrote "Is man an ape or an angel? . . . I am on the side of the angels. I repudiate with indignation and abhorrence these new-fangled theories."

Nevertheless there is no doctrinal conflict between Christianity and neo-Darwinism properly understood. It is possible to be both a convinced Christian and an orthodox evolutionist: Christians believe that God is active through natural as well as supernatural causes. God can be described as Creator and at the same time accepted as at work through mutation, selection, and so on, in exactly the same way as a painter can design and execute a picture, which is nonetheless also describable as a pattern of molecules.

Humans are described in the Bible as emerging from a similar process as the other organisms, but then changed "into God's image." (The Hebrew word used for the creation of matter, life and humankind implies a special creation; the usual word for God's work in Genesis 1 means "molding," as a potter molds clay.) There is no reason to believe that the process of "spiritual" creation would alter *Homo sapiens* genetically, physiologically or anatomically; Christians believe that it makes the human species distinct from the rest of animals and gives a special relationship to God. But as a biological species we are evolutionarily related to the apes.

A man "made in God's image" can "fall" from this state. This removes one of the Christian's main problems about evolution: the belief that acceptance of evolution does not automatically imply that man is still "on the way up," and hence improving. There is no anthropological or theological evidence for any general moral progress of man; and the political dream of improving behavior by improving the environment has seldom if ever worked.

T. H. Huxley, "Darwin's bulldog," was very condemnatory about the influence of religion on science, but he was realistic about the nature of man: "It is the secret of the superiority of the best theological teachers to the majority of their opponents, that they substantially recognize the realities of things. The doctrine of predestination; of original sin; of the innate depravity of man appear to me to be vastly nearer the truth than the 'liberal' popular illusions that babies are all born good and that the example of a corrupt society is responsible for their failure to remain so; and that it is given to everybody to reach the ethical ideal if only he will try."   RJB

However, the final collapse of the extreme neutralist position came from the recognition that many (and perhaps most) variants change the properties of the proteins in which they are substituted, and are thus potentially subject to selection. This has been shown by studies on biochemical properties *in vitro* by correlating variant distribution with environmental variables (particularly temperature), and on changes of allele frequencies in time and space. The firm conclusion is that allele frequencies of all sorts are liable to fine adjustment by natural selection.

The theoretical arguments of Muller and Haldane can in retrospect be seen to be rather naive. Both men effectively thought of each gene acting independently on its carrier. This is patently not true: the entire phenotype is selected, not independent aspects of it. Furthermore, selection pressures change in time and space, depending on both the physical and the ecological environment. (This is not Lamarckism: gene transmission is not affected by the environment, although gene expression is.) The genetic composition of a population or species cannot be separated from its history. Indeed, genetic composition is in some ways a record of the impacts and vicissitudes of past environments. Some alleles will respond to current environmental pressures, but others will merely reflect past events. For example, human blood-group frequencies were probably determined by the major epidemic diseases of the past, as individuals died or survived depending on their resistance to particular diseases. Nowadays blood-group differences seem to have little effect on survival.

Judged rather harshly, much of the support for the neutral mutation point of view ("non-Darwinian evolution" as it has been called) came from the elegance of the mathematical arguments rather than their empirical correctness as such. This does not mean that the neutralist criticisms of neo-Darwinism were completely wrong. Some mutations *will* have no effect on fitness for much of the life of their carriers; the crucial point is that the neutrality of such alleles is unlikely to persist indefinitely.

The neutralist–selectionist controversy had two effects. First, it broke the stranglehold of theoreticians on evolutionary biology which had developed during the 1950s, and which threatened to turn the subject into a branch of applied mathematics. Secondly, biologists were recalled to biological phenomena in the widest sense (biochemical, immunological, behavioral and physiological, as well as ecological and morphological), and saw the need to resist the temptation to seek all their answers from molecular biology.

Continuing Controversies

Modern evolutionary theory is the result of at least four syntheses: (1) Darwin's own debate with morphologists, taxonomists and naturalists in the *Origin of Species* itself. (2) The disputes between biometricians and Mendelists in the years following the rediscovery of Mendel's results in 1900. (3) The main neo-Darwinian synthesis between paleontologists and geneticists in the 1930s. (4) The neutralist–selectionist controversy of the 1960s.

Continuing debates about evolution are not about the *fact* of evolution (except among "creationists"), but about the adequacy of the Darwinian mechanism of adaptation through natural selection as a sufficient cause for change.

Punctuated Equilibrium

In 1972 Niles Eldredge and Stephen Jay Gould pointed out that many evolutionary lineages begin suddenly in the fossil record with the appearance of a new species, while other fossil lines remain unchanged for millions of years. "Punctuations" alternate with periods of "stasis" or equilibrium. This is not the place to go into the detail of this debate, except to say that suddenness in paleontological time is very different from suddenness in genetical time; the splitting of a lineage can take place very rapidly (in tens rather than thousands of generations) and be produced by a small number of gene differences (as well as by polyploidy, which is the classical method of "instant speciation").

The problem with "punctuated equilibrium" is that critics of Darwinism have equated it with the theory that new species

originate from mutations, as postulated by de Vries and Richard Goldschmidt, and so have assumed it is distinct from the speciation process represented by Darwin in the phrase *natura non facit saltum* ("nature does not make jumps"). This assumption is false: Darwin himself was quite clear that evolutionary rates vary. He is explicit in the *Origin of Species* that "the periods during which species have been undergoing modification, though very long as measured by years, have probably been short in comparison with the periods during which these same species remained without undergoing any change," and "I do not suppose that the process [speciation] ... goes on continuously; it is far more probable that each form remains for long periods unaltered, and then again undergoes modification." G. G. Simpson, in his classical exposition of the neo-Darwinian synthesis (*Tempo and Mode in Evolution*, 1944), stated: "The pattern of steplike evolution that has the appearance of successive structural steps, rather than direct sequent phyletic transitions, is a peculiarity of paleontological data more nearly universal than true rectilinearity." There is no basic conflict between punctuated equilibrium and Darwinian evolution properly understood.

The point that evolutionary rates vary very widely is pertinent to objections that some evolutionary events are so unlikely that they could never have happened by chance. For example, the British astronomer Sir Fred Hoyle has argued that the idea that life originated by random shuffling of molecules is "as ridiculous and improbable as the proposition that a tornado blowing through a junk yard may assemble a Boeing 747." He calculated that the likelihood of life beginning in this way is one in ten to the power of 40,000—the chance that 2,000 enzyme molecules will be formed simultaneously from their 20 component amino acids on a single specified occasion. But this is not the correct calculation; the relevant chance is of some far simpler self-replicating system, capable of development by natural selection, being formed at any place on earth, and at any time within a period of 100 million years. We cannot

The Scopes trial (10–21 July, 1925). John Scopes LEFT, a teacher, was convicted of teaching Darwinism, thus breaking a Tennessee law which made it unlawful to teach any doctrine denying divine creation. The law was repealed in 1967.

calculate this probability, since we know neither the nature of the hypothetical self-replicating system nor the composition of the "primeval soup" in which it arose. The origin of life was obviously a rare event, but there is no reason to think that it is as extraordinary or unlikely as Hoyle calculated.

Hoyle's argument is somewhat similar to the often repeated claim that the evolution of complex functioning organs like the human eye or ear by "blind" selection acting on random mutations is as unlikely as a monkey producing the plays of Shakespeare by hitting the keys of a typewriter at random. This completely misunderstands the role and power of natural selection in producing adaptation. R. A. Fisher has commented that "it was Darwin's chief contribution to have brought to light a process by which contingencies *a priori* improbable are given, in the process of time, an increasing probability, until it is their nonoccurrence which becomes highly improbable."

Cladism

The classifying of animals and plants (taxonomy) has traditionally been a subjective exercise, controlled by the general agreement between specialists as to the reality of particular species. For many years biologists have tried to make classifications more exact. They have used more characters (including chemical and behavioral ones) than the usual morphological criteria, and have increasingly used quantitative multivariate techniques. One particular refinement of this trend is cladism, a method of systematics derived from the work of a German taxonomist, Willi Hennig, and based on a book by him published in English in 1966 (*Phylogenetic Systematics*). Cladism has three axioms: (1) Features shared by organisms manifest a hierarchical pattern in nature. (2) This hierarchical pattern may be economically expressed by means of branching diagrams (cladograms). (3) The nodes (branching points) in a cladogram symbolize the homologies shared by the organisms linked by the node, so that a cladogram is congruent with a classification.

The cladistic method imposes a discipline on taxonomists, but in the process it has become divorced from evolutionary biology, and especially the working hypothesis of most systematists that they are largely reconstructing phylogeny (evolutionary ancestry) in their classifications. Indeed, some cladists explicitly deny that they are concerned with evolution; they claim that they are concerned solely with "pattern" in nature, in ways reminiscent of pre-Darwinian romantics such as Goethe and Geoffroy Saint-Hilaire. In this respect they have departed from Hennig's own method, which was consciously phylogenetic; ironically, the major theoretical contribution of Hennig to classification was in distinguishing between primitive and derived homologies, which is necessary before an evolutionary phylogeny can be constructed.

The cladism controversy has obscured the fundamental principle that while there can be only one true phylogeny for a group (though we may not be sure what it is), classification is more arbitrary, because the rules of classification are made by people for their own convenience. In the words of one of its main protagonists, cladism is "merely rediscovering preevolutionary systematics; or if not rediscovering it, fleshing it out." Cladistics is not antievolutionary; it is simply irrelevant to a discussion of evolutionary mechanism.

Neo-Lamarckism

The simplistic idea that characters acquired during life could be inherited by the offspring, as advocated by Jean-Baptiste Lamarck in the early 1800s, had to be abandoned once the physical basis of heredity was worked out. Nevertheless, there have been many attempts to show that some form of Lamarckian inheritance occurs in nature. None has been confirmed.

The most infamous "neo-Lamarckian" episode occurred in Russia between 1948 and 1964. The leader of this movement, T. D. Lysenko, claimed to be able to change the inherited properties of crop plants by grafting or seed treatment, and thereby permanently to improve yield by environmental manipulation. No one outside the USSR was able to repeat these results, and eventually Lysenko was discredited within his own country. Traditional genetics (or "Mendelist–Morganism" as it was called) was suppressed during Lysenko's ascendancy, and many of its practitioners imprisoned. The fact that genes and chromosomes carry heritable material was denied, and replaced by the notion that heredity was a general property of living matter. Lysenko received considerable support from the Soviet authorities because his claims coincided with Marxist ideas of the possibility of change. Controversies about the inheritance of intelligence are in some ways a continuance of the Lamarckian debate.

Creationism

The most persistent challenge to neo-Darwinism comes from certain Christian groups, and also some Islamic sects. Darwin's argument that the present state of biological diversity has arisen by an entirely mechanistic process was anathema to many, and has remained so to some. The prosecution in 1925 of John Scopes for teaching Darwinism in a Tennessee school is well-known. For 40 years after that, little was heard of the creationist controversy. It surfaced again in 1972 when the California State Board of Education required that creationism be accorded "equal time" with evolution in state schools.

Creationists insist that the human race was specially created by God, and has no genetic relationship to any other species. They extrapolate from this to deny that any significant evolution has occurred. Many disputes are taken out of their contexts and used by creationists as support for their contention that evolution and evolutionary theory are nothing more than dubious hypothesis. For example, it is claimed that all the phyla except the vertebrates are present in the earliest fossiliferous rocks (ignoring Precambrian fossils); that the theory of punctuated equilibrium is inconsistent with neo-Darwinism; that all mutations are harmful; that the most important hereditary determinants are in cell membranes and not nuclear DNA; that no species transitions are known (except the trivial case of polyploidy); that evolution is contrary to the Second Law of Thermodynamics (which only applies in a system receiving no energy, but the earth is continually receiving energy from the sun); that all rock-dating techniques are unreliable; and so on. Creationist literature is a sadly distorted microcosm of the whole neo-Darwinian debate. The tragedy of the creationist position is that it begins from the axiom that evolution is unthinkable, and hence it preempts rational discussion on the issue.   RJB

Bibliography

The literature on evolution and paleontology is vast, ranging from general books on the subject to specialized monographs on peripheral topics such as geophysics or behavioral ecology, and from the elementary to the propagandist. A complete bibliography would be longer than this book. Hence the list that follows is selective, compiled with the aim of including historically important books and the more significant modern ones. It should be regarded as a guide to further reading, rather than a comprehensive reference source.

Paleontology and Historical Geology
Eicher, D.L., McAlester, A.L. and Rottman, M.L. (1984) *The History of the Earth's Crust*, Prentice-Hall, Englewood Cliffs, New Jersey.
Eldredge, N. and Cracraft, T. (1980) *Phylogenetic Patterns and the Evolutionary Process*, Columbia University Press, New York.
Fortey, R.A. (1982) *Fossils: the Key to the Past*, Heinemann, London.
Hallam, A. (ed) (1977) *Patterns of Evolution as Illustrated by the Fossil Record*, Elsevier, Amsterdam.
Lane, N.G. (1978) *Life of the Past*, Merrill, Westerville.
Nitecki, M.H. (1984) *Extinctions*, University of Chicago Press, Chicago.
Paul, C. (1980) *The Natural History of Fossils*, Weidenfeld and Nicolson, London.
Raup, D.M. and Jablonski, D. (eds) (1986) *Pattern and Process in the History of Life*, Springer-Verlag, Berlin.
Raup, D.M. and Stanley, S.M. (1978) *Principles of Palaeontology* (2nd edn), Freeman, San Francisco.
Stanley, S.M. (1986) *Earth and Life through Time*, Freeman, San Francisco.

Fossil Groups
Charig, A.L. (1979) *A New Look at the Dinosaurs*, Heinemann, London.
Clarkson, E.N.K. (1979) *Invertebrate Palaeontology and Evolution*, Allen and Unwin, London.
Glaessner, M.F. (1984) *The Dawn of Animal Life*, Cambridge University Press, Cambridge.
Kemp, T.S. (1982) *Mammal-like Reptiles and the Origin of Mammals*, Academic Press, London.
Savage, D.E. and Russell, D.E. (1983) *Mammalian Palaeofaunas of the World*, Addison-Wesley, Reading, Mass.
Taylor, T.N. (1981) *Palaeobotany: An Introduction to Fossil Plant Biology*, McGraw-Hill, London.

Darwin
No serious student of evolution can ignore Charles Darwin's own writings, notably of course his *On the Origin of Species by Means of Natural Selection, or the preservation of favoured races in the struggle for life*, first published by John Murray in 1859, and revised five times. This is available in a number of cheap editions. Other key books by or about Darwin include:
Appleman, P. (ed) (1970) *Darwin*, Norton, New York.
Atkins, H. (1974) *Down, the Home of the Darwins*, Royal College of Surgeons, London.
Barrett, P.H. (ed) (1977) *The Collected Papers of Charles Darwin*, University of Chicago Press, Chicago.
Clark, R.W. (1984) *The Survival of Charles Darwin*, Weidenfeld and Nicolson, London.
Darwin, C. and Wallace, A.R. (1958) *Evolution by Natural Selection: a centenary commemorative volume*, Cambridge University Press, Cambridge.
Freeman, R.B. (1978) *The Works of Charles Darwin: a Companion*, Dawson-Archon, Folkstone.
Huxley, J. and Kettlewell, H.B.D. (1965) *Darwin and His World*, Thames and Hudson, London.
Keynes, R.D. (ed) (1979) *The Beagle Record*, Cambridge University Press, Cambridge.
Moorehead, A. (1969) *Darwin and the Beagle*, Hamish Hamilton, London.
Raverat, G. (1952) *Period Piece*, Faber, London.
Stauffer, R.C. (ed) (1975) *Charles Darwin's Natural Selection*, Cambridge University Press, Cambridge.

Histories of Evolution
Bowler, P. (1983) *The Eclipse of Darwinism*, Johns Hopkins University Press, Baltimore.
Carter, G.S. (1957) *A Hundred Years of Evolution*, Sidgwick and Jackson, London.
Clodd, E. (1897) *Pioneers of Evolution from Thales to Huxley*, Grant Richards, London.
Gillespie, N.C. (1979) *Charles Darwin and the Problem of Creation*, University of Chicago Press, Chicago and London.
Hull, D.L. (1973) *Darwin and His Critics*, Harvard University Press, Cambridge, Mass.
Irvine, W. (1955) *Apes, Angels and Victorians*, Weidenfeld and Nicolson, London.
Kellogg, V.L. (1907) *Darwinism Today*, Bell, New York and London.
Mayr, E. (1982) *The Growth of Biological Thought*, Belknap, Cambridge, Mass and London.
Moore, J.R. (1979) *The Post-Darwinian Controversies*, Cambridge University Press, Cambridge.
Ospovat, D. (1981) *The Development of Darwin's Theory. Natural History, Natural Theology and Natural Selection 1838–1859*, Cambridge University Press, Cambridge.
Provine, W.B. (1971) *The Origins of Theoretical Population Genetics*, Chicago University Press, Chicago and London.
Seward, A.C. (ed) (1909) *Darwin and Modern Science*, Cambridge University Press, Cambridge.

The Neo-Darwinian Synthesis
Berry, R.J. (1982) *Neo-Darwinism*, Edward Arnold, London.
De Beer, G.R. (1930) *Embryology and Evolution*, Clarendon, Oxford. (Revised as *Embryos and Ancestors*, 1940, 3rd edn 1958.)
Dobzhansky, T. (1937, 3rd edn 1951) *Genetics and the Origin of Species*, Columbia University Press, New York. (Revised as *Genetics of the Evolutionary Process*, 1970.)

Fisher, R.A. (1930, 2nd edn 1958) *The Genetical Theory of Natural Selection*, Clarendon, Oxford. (2nd edn published by Dover, New York.)
Ford, E.B. (1931, 8th edn 1965) *Mendelism and Evolution*, Methuen, London.
Goldschmidt, R.B. (1940) *The Material Basis of Evolution*, Yale University Press, New Haven.
Haldane, J.B.S. (1932) *Causes of Evolution*, Longman, London.
Huxley, J.S. (ed) (1940) *The New Systematics*, Oxford University Press, London.
Huxley, J.S. (1942) *Evolution, the Modern Synthesis*, Allen and Unwin, London.
Mayr, E. (1942) *Systematics and the Origin of Species*, Columbia University Press, New York. (Revised as *Animal Species and Evolution*, 1963, Harvard, Cambridge, Mass and Oxford University Press, London.)
Mayr, E. and Provine W.B. (1980) *The Evolutionary Synthesis*, Harvard University Press, Cambridge, Mass.
Rensch, B. (1959) *Evolution above the Species Level*, Columbia University Press, New York.
Simpson, G.G. (1944) *Tempo and Mode in Evolution*, Columbia University Press, New York. (Revised as *Major Features of Evolution*, 1953.)
Stebbins, G.L. (1950) *Variation and Evolution in Plants*, Columbia University Press, New York.

General Works on Evolution
Bendall, D.S. (ed) (1983) *Evolution from Molecules to Men*, Cambridge University Press, Cambridge.
Bowler, P.J. (1984) *Evolution. The History of an Idea*, University of California Press, Berkeley.
Cain, A.J. (1954) *Animal Species and their Evolution*, Hutchinson, London.
Calow, P (1983) *Evolutionary Principles*, Blackie, Glasgow.
Carter, G.S. (1951) *Animal Evolution*, Sidgwick and Jackson, London.
Chapman, R.G. and Duval, C.T. (eds) (1982) *Charles Darwin. A Centennial Commemorative*, Nova Pacifica, Wellington, New Zealand.
Cherfas, J. (ed) (1982) *Darwin up to Date*, New Scientist, London.
Dobzhansky, T., Ayala, F.J., Stebbins, G.L. and Valentine, J.W. (1977) *Evolution*, Freeman, San Francisco.
Gould, S.J. (1977) *Ever Since Darwin*, Norton, New York.
Gould, S.J. (1980) *The Panda's Thumb*, Norton, New York and London.
Gould, S.J. (1983) *Hen's Teeth and Horse's Toes*, Norton, New York.
Grant, V.M. (1963) *The Origin of Adaptations*, Columbia University Press, New York.
Grant, V.M. (1977) *Organismic Evolution*, Freeman, San Francisco.
Grene, M. (ed) (1983) *Dimensions of Darwinism*, Cambridge University Press, Cambridge.
Huxley, J. (1953) *Evolution in Action*, Chatto and Windus, London.
Huxley, J.S., Hardy, A.C. and Ford, E.B. (eds) (1954) *Evolution as a Process*, Allen and Unwin, London.
Maynard Smith, J. (1958, 3rd edn 1975) *The Theory of Evolution*, Penguin, Harmondsworth.
Maynard Smith, J. (ed) (1982) *Evolution Now: A Century After Darwin*, Macmillan, London.
Patterson, C. (1978) *Evolution*, Routledge & Kegan Paul, London.
Sober, E. (ed) (1984) *Conceptual Issues in Evolutionary Biology*, M.I.T., Cambridge, Mass.

Williams, G.C. (1966) *Adaptation and Natural Selection*, Princeton University Press, Princeton.
Wright, S. (1968–1978) *Evolution and the Genetics of Populations*, 4 vols, Chicago University Press, Chicago.

Particular Aspects of Evolution
Aiello, L. (1982) *Discovering the Origins of Man*, Longman, London.
Ayala, F.J. (ed) (1976) *Molecular Evolution*, Sinauer, Sunderland, Mass.
Bennett, J.H. (ed) (1965) *Experiments in Plant Hybridisation. Gregor Mendel*, Oliver and Boyd, Edinburgh and London.
Berry, R.J. (1977) *Inheritance and Natural History*, Collins New Naturalist, London.
Berry, R.J. (ed) (1984) *Evolution in the Galapagos Islands*, Academic Press, London.
Bonner, J.T. (ed) (1981) *Evolution and Development*, Springer-Verlag, Berlin.
Brown, J.L. (1975) *The Evolution of Behavior*, Norton, New York.
Browne, J. (1983) *The Secular Ark*, Yale University Press, New Haven.
Cavalli-Sforza, L.L. and Wodmer, W.F. (1971) *The Genetics of Human Populations*, Freeman, San Francisco.
Dawkins, R. (1976) *The Selfish Gene*, Oxford University Press, Oxford.
Flew, A. (1984) *Darwinian Evolution*, Paladin, London.
Forey, P. (ed) (1981) *The Evolving Biosphere*, British Museum (Natural History), London.
Frankel, O.H. and Soulé, M.E. (1981) *Conservation and Evolution*, Cambridge University Press, Cambridge.
Futuyama, D. (1979) *Evolutionary Biology*, Sinauer, Sunderland, Mass.
Futuyama, D.J. and Slatkin, M. (eds) (1983) *Coevolution*, Sinauer, Sunderland, Mass.
Gould, S.J. (1977) *Ontogeny and Phylogeny*, Harvard University Press, Cambridge, Mass and London.
Grant, V.M. (1964) *Architecture of the Germplasm*, Wiley, New York and London.
Grayson, D.K. (1983) *The Establishment of Human Antiquity*, Academic Press, New York.
Kettlewell, H.B.D. (1973) *The Evolution of Melanism*, Clarendon, London.
Kimwa, M. (1983) *The Neutral Theory of Molecular Evolution*, Cambridge University Press, Cambridge.
Leakey, R.E. (1981) *The Making of Mankind*, Michael Joseph, London.
Leakey, R.E. and Lewin, R. (1977) *Origins: What New Discoveries Reveal about the Emergence of Our Species and its Possible Future*, Macdonald and Janes, London.
Lewontin, R.C. (1974) *Genetic Basis of Evolutionary Change*, Columbia University Press, New York.
Mason, I.L. (ed) (1984) *Evolution of Domesticated Animals*, Longman, London and New York.
Mather, K. (1974) *Genetical Structure of Populations*, Chapman and Hall, London.
Milkman, R. (ed) (1982) *Perspectives on Evolution*, Sinauer, Sunderland, Mass.
Nei, M. and Koehn, R.K. (eds) (1983) *Evolution of Genes and Proteins*, Sinauer, Sunderland, Mass.
Parkin, D.T. (1979) *Introduction to Evolutionary Genetics*, Arnold, London.
Raff, R.A. and Kaufman, T.C. (1983) *Embryos, Genes and Evolution*, Macmillan, New York.
Schopf, T.J.M. (ed) (1972) *Models in Paleobiology*, Freeman, San Francisco.
Scientific American (1978) *Evolution*, Freeman, San Francisco.

Sheppard, P.M. (1958, 4th edn 1975) *Natural Selection and Heredity*, Hutchinson, London.
Sibley, R.M. and Smith, R.H. (eds) (1985) *Behavioural Ecology: Ecological Consequences of Adaptive Behaviour*, Blackwell, Oxford.
Stanley, S.M. (1979) *Macroevolution. Pattern and Process*, Freeman, San Francisco.
Stanley, S.M. (1981) *The New Evolutionary Timetable*, Basic, New York.

Waddington, C.H. (1957) *The Strategy of the Genes*, Allen and Unwin, London.
Weiner, J.S. (1971) *Man's Natural History*, Weidenfeld and Nicolson, London.
Wickler, W. (1968) *Mimicry in Plants and Animals*, McGraw-Hill, New York.
Williamson, M. (1981) *Island Populations*, Oxford University Press, Oxford.

Religious Questions
Berry, R.J. (1975) *Adam and the Ape*, Falcon, London.

Divant, J. (ed) (1985) *Darwinism and Divinity*, Blackwell, Oxford.
Futuyama, F.J. (1983) *Science on Trial*, Pantheon, New York.
Godfrey, L.R. (ed) (1983) *Scientists Confront Creationism*, Norton, New York.
Kitcher, P. (1983) *Abusing Science*, Open University Press, Milton Keynes.
Lack, D. (1957) *Evolutionary Theory and Christian Belief. The unresolved conflict*, Methuen, London.

Morris, H.M. (ed) (1974) *Scientific Creationism*, Creation-Life, San Diego, California.
Peacock, A.R. (1979) *Creation and the World of Science*, Clarendon, Oxford.
Ruse, M. (1982) *Darwinism Defended*, Addison-Wesley, Reading, Mass.
Whitcomb, J.C. and Morris, H.M. (1961) *The Genesis Flood*, Presbyterian and Reformed, Philadelphia.

Picture Acknowledgements

Key: *t* top. *b* bottom. *c* center. *l* left. *r* right.
Abbreviations: ANT Australasian Nature Transparencies. BCL Bruce Coleman Ltd. BHPL BBC Hulton Picture Library. MEPL Mary Evans Picture Library. NHPA Natural History Photographic Agency. OSF Oxford Scientific Films. PEP Planet Earth Pictures. SAL Survival Anglia Ltd. SPL Science Photo Library.

5 Mansell Collection. 6–7 SPL. 7*b* Zefa/W.F. Davidson. 7*t* SPL/Dr. Ann Smith. 7*c* Zefa/H. Helmlinger. 10 Dr. R. Goldring. 11 Zefa/C. Maher. 13*t* M.R. Walter. 13*b* R.J. Jenkins. 15*t* S. Conway-Morris. 15*b* Travel Photo International. 21 Geoscience Features. 34 Russell Lamb. 36 Dr. Jens Lorenz Franzen. 48 Imitor. 50–51 Andrew Laurie. 52*r*, 54*r*, 54*l*, 55*b* BHPL. 55*t*, 56 MEPL. 57*t*, 57*b* BHPL. 57*c*, 58*tl* Mansell Collection. 58*cl*, 58*bl* BHPL. 58*r* MEPL. 60 BHPL. 61*tl* Mansell Collection. 61 inset MEPL. 64–65, 66–67 Sinclair Stammers. 67*b* Biophoto Associates. 70–71 OSF/G.I. Bernard. 74 Topham Picture Library. 75*t* Mansell Collection. 75*b* British Museum (Natural History). 76*t* PEP/W. Deas. 76–77 Biofotos/H. Angel. 77*t* ANT/A. Dennis. 77*c* PEP/P. Scoones. 77*b* G. Bateman. 78–79 Frank Lane Agency/S. Jonasson. 80 SPL/J. Burgess. 81*t* Sio Photo. 81*b* S.W. Fox. 83 MEPL. 84–85 SAL/J.&D. Bartlett. 84*b* PEP/J. Scott. 85*b* C.A. Henley. 88 Premaphotos.

89*l* Mansell Collection. 89*r* BHPL. 93 Ardea/R.J.C. Blewitt. 94 PEP/P. Scoones. 94–95 Frank Lane Agency/G. Moon. 95*b* Premaphotos. 98 Biofotos/H. Angel. 99*t* Premaphotos. 99 inset SPL/V. Fleming. 99*b* C.A. Henley. 102–103 SPL/Dr. Gopal Murti. 105 BHPL. 106, 107 SPL. 112 Ardea/P. Morris. 113 Dr. G. Mazza. 114*t* BCL/R. Coleman. 114*b* BCL/E. Crichton. 118 BCL/J.&D. Bartlett. 120*l*, 120–121 R. Tidman. 120*r* E.&D. Hosking. 122–123 Frank Lane Agency/C. Carvalho. 123 Leonard Lee Rue III. 126–127 Illustrated London News. 127*t* Dr. L.M. Cook. 128–129 Zefa/Kappel Meyer. 130 Zefa/V. Englebert. 131*tl*, 131*tr* Camerapix Hutchison Library. 131*tc* BCL/M. Fogden. 131*bl* NHPA/P. Johnson. 131*br* Zefa/R. Smith. 136 NHPA/P. Johnson. 137 B.&C. Alexander. 139 MEPL. 142,144 BHPL.

Artwork

Abbreviations: SD Simon Driver. OI Oxford Illustrators Limited. ML Michael Long. DO Denys Ovenden.

2,3 ML. 4 SD. 7 OI. 10,11 SD. 16,17,18,19,20*b*,22,23,26,27,28,29,31,32,33,34,39,40, 41,42,43,44,45,46,47 ML. 48,63,66,67 OI. 68,69 Richad Orr. 70 Mick Loates. 71 SD. 72,73*b* Richard Orr. 73*c* SD. 81,83 OI. 84,85 SD. 86,87 DO. 90,91 OI. 96,97 Richard Lewington. 100,101 Graham Allen. 111*b*,113,115 OI. 116,117 Trevor Boyer. 119 OI. 120 Nicholas Rous. 123 OI. 124 Richard Lewington. 125 Trevor Boyer. 127*t* Nicholas Rous. 127*b*,130,131 OI. 132, 133*tr*,134,135 Richard Hook. 133*tl* OI. Other artwork: Equinox (Oxford) Limited.

GLOSSARY

Adaptation anything which enhances an organism's chances of survival in the environment in which it lives. Adaptations may be GENETIC, produced by evolution, and hence not alterable within the animal's lifetime, or they may be PHENOTYPIC, produced by adjustment on the part of the individual and reversible within its lifetime.

Adaptive radiation the evolutionary diversification of a TAXON into a number of subsidiary groups adapted to more restrictive modes of life. This occurs typically over a relatively short period of time.

Adh alcohol dehydrogenase; the enzyme catalyzing the oxidation of ethanol to acetaldehyde.

Age and area this hypothesis proposes that the area occupied by a species is proportional to its evolutionary age.

Allele any of the different forms of a GENE that occupies a particular locus.

Allometric growth differential growth of body parts, leading to a change of shape or proportion.

Allopatric speciation see SPECIATION.

Altruism behavior that brings advantage to others than those who perform it. The term is usually taken to include a reduction in the performer's own FITNESS.

Ammonoid pertaining to the Ammonoidea, an order of extinct cephalopod mollusks of the subclass Tetrabranchia; they are important as "index" fossils.

Amnoitic eggs eggs with an amnion, a thin extra-embryonic membrane which forms a closed sac around the embryo in birds, reptiles and mammals (the Amniota).

Analogy similarity of structure, function or behavior due to CONVERGENT EVOLUTION as opposed to common ancestry (HOMOLOGY).

Angiosperms plants with seeds that are enclosed in an ovary; flowering plants.

Arachnid member of the class Arachnida, phylum Chelicerata. Arachnids are mostly terrestrial and have four pairs of thoracic appendages. Examples include spiders, scorpions and mites.

Arthropod member of a group of phyla known as "Arthropoda." Arthropods typically have a segmented body, hard integument and paired jointed limbs; they include the crustaceans and the insects. Some authorities regard the Arthropoda as a distinct phylum, rather than several phyla.

Asexual reproduction reproduction that does not involve FERTILIZATION and which is devoid of any sexual process such as MEIOSIS.

Asymmetry lack of symmetry; skewness. This often refers to the positions of parts of the body with respect to imaginary axes. Cf BILATERAL SYMMETRY.

ATP adenosine triphosphate. This is a comparatively unstable substance, readily breaking down to adenosine diphosphate (ADP), and releasing a large amount of energy in the process. ATP is the chief energy currency in the metabolism of organisms.

Bacterium extremely small and simple PROKARYOTIC organism.

Base any chemical capable of accepting a proton from another substance. Purines and pyrimidines are classes of organic bases found in nucleic acids such as DNA. The pairing of complementary purine and pyrimidine bases joins the component strands of the DNA double helix.

Benthos (benthic) the bottom layer of the aquatic environment.

Beringia Miocene land mass incorporating NW America and NE Eurasia and the present position of the Bering Straits.

Bilateral symmetry a bilaterally symmetrical animal can be halved in one plane to give two halves which are mirror images of each other. Most multicellular animals display this form of symmetry, and it is generally associated with a mobile, free-living way of life.

Binary fission a form of ASEXUAL REPRODUCTION of a cell in which the nucleus divides, and then the CYTOPLASM splits into two approximately equal parts.

Biogenic law the theory that the embryonic development of an individual approximately retraces its species' evolutionary history (Haeckel's Law: "ontogeny recapitulates phylogeny"). This idea is now thought to be erroneous.

Biomass a measure of the abundance of a life form in terms of its mass, either absolute or per unit area.

Biome a major regional ecological community characterized by a distinctive flora and fauna.

Biometry the application of statistical methods to the analysis of biological data.

Biosystem the entire biological and physical content of a particular area.

Biota the collective FAUNA and FLORA of a given region or area.

Body fossil the fossil remains of an organism (as opposed to imprint, trail or trace).

Body plan the generalized (archetypal) body structure of a major TAXON.

Boulder clay (till) the unstratified material deposited by glaciers and ice sheets.

Calcareous composed of, or rich in, calcium carbonate.

Cambrian period the first geological period of the PALEOZOIC ERA (about 590-505 million years ago). This period was marked by shallow seas and a warm climate. There are records of trilobites and brachiopods and the first remains of vertebrates are found at the end of the period.

Carboniferous period a PALEOZOIC geological period (about 360-286 million years ago), subdivided into the Mississippian and Pennsylvanian. This period was marked by extensive lowland forests and swamps which were the beginning of the great coal deposits, and by the considerable radiation of the amphibians.

Carnivore any meat-eating organism; alternatively, a member of the mammalian order Carnivora, many of whose members are carnivores.

Cenozoic era the most recent geological era, consisting of the TERTIARY and QUATERNARY periods, about 65 million years ago to the present. This is sometimes called the "age of mammals."

Chert a silica-based rock often composed of fossil microorganisms, especially radiolaria.

Chloroplast the chlorophyll-containing body in plant cells in which PHOTOSYNTHESIS takes place.

Chordate a member of the phylum Chordata which comprises all animals which at any stage in their development possess a notochord. They include all the vertebrates as well as the lower groups Urochordata (tunicates) and Cephalochordata (*Amphioxus*, the lancelet).

Chromosome a microscopic, thread-like body that is composed of DNA and PROTEIN and becomes visible in the nucleus of the cell at the time of cell division. Chromosomes carry the GENES and are normally constant in number within a species but may vary in number between even quite closely related species.

Cladistics (cladism) a method of classification using solely the criterion of recency of common ancestor to group organisms: two taxa are more closely related to each other than to a third if they share a more recent common ancestor. The product of a cladistic classification is a **cladogram**, which is a branching diagram representing the relationships between taxa.

Class a rank in the taxonomic hierarchy coming between PHYLUM or division and ORDER. In zoology, classes typically end with an -a (eg Vertebrata), and, in botany, with -opsida.

Clay a fine-grained sediment composed mostly of clay minerals (hydrous aluminum silicates) and other material less than 4 microns in diameter.

Cline continuous variation in a character through a series of contiguous populations.

Clone a group of organisms derived by ASEXUAL or vegetative reproduction from a single individual.

Coacervate an aggregation of organic molecules, resulting in particulate matter.

Coadaptation the evolution of mutually advantageous traits in interacting species.

Coccoliths the calcareous plates enclosing the bodies of some marine planktonic protozoa.

Codon a set of three nucleotide bases required to specify one amino acid; the unit of genetic coding.

Coelenterate a member of the group Coelenterata, consisting of the phyla Cnidaria (anemones, jellyfish and corals) and Ctenophora (sea gooseberries). These animals usually have a radially symmetrical sac-like body with a mouth at one end, often surrounded by a ring of stinging tentacles.

Coevolution the interdependent evolution of two interacting species.

Community all the living organisms in any one place, which interact with each other in some way.

Continental shelf the shallow sloping seabed around the edge of a continent.

Convergent evolution the independent acquisition of similar characters in evolution, as opposed to the possession of similarities by virtue of descent from a common ancestor.

Creationism the belief that all life was created in its present form and has not subsequently undergone change.

Cretaceous period a geological period in the MESOZOIC ERA (about 144-65 million years ago). This was a period of extensive mountain formation and climatic cooling and marked the start of ANGIOSPERM dominance.

Crypsis concealment by the use of camouflaged coloring or markings and/or by specialized behavior.

Cynodont a member of the extinct reptile group, Cynodontia, from the TRIASSIC PERIOD. These had dog-like teeth and many mammalian characters.

Cytology the study of the structure and function of cells.

Cytoplasm the material of which a cell, excluding its nucleus, is made.

Darwin a unit measuring rate of evolutionary change; it represents an increase or decrease in any given character by a factor of 2.7 per million years.

Deoxyribose the sugar component of a molecule of DNA, the genetic material.

Detritus organic debris derived from the decomposition of plant and animal remains.

Devonian period a geological period in the PALEOZOIC ERA (about 408-360 million years ago). This was marked by cool climates and the appearance of the first forests, winged insects and tetrapods.

Diagenesis the physical and chemical process involved in rock formation after the initial sedimentation.

Diploid having the homologous sets of CHROMOSOMES in the cell nucleus; cf HAPLOID.

DNA deoxyribonucleic acid, the genetic material in the nuclei of cells. This consists of double helical polymer of molecules (NUCLEOTIDES) containing a sugar (DEOXYRIBOSE), a purine or pyrimidine BASE and a phosphate group. The sequence of the different pyrimidine or purine bases along the polymer is the basis of the genetic code.

Dolomite a carbonate mineral. The term is also used to denote rock with a high ratio of magnesium to calcium carbonate.

Dominant genetic: a trait that appears when the gene carrying it is present on only one of two chromosomes in a pair, and has the same expression when the gene (ALLELE) is present on both chromosomes (ie in the homozygous condition). *Also*, Behavioral/ecological: physical domination of others either by individuals within a species or by a species within a community.

Ecology the study of the interaction between living organisms and their environment.

Ecosystem a system consisting of the interaction of a community of organisms with their environment.

Ediacaran fauna the animals of the last PRECAMBRIAN geological period (about 700-590 million years ago) which provide the earliest clear insight into early animal history.

Embryology the study of the formation and early development of living organisms.

Endemic a group of species which is restricted to, and native to, a particular geographical area.

Entelechy evolution in one direction over a considerable period of time. This has been taken to imply, incorrectly, that there is an internal evolutionary drive within the organism.

Entropy a measure of the disorder of a system: a highly disordered system has a high entropy.

Enzyme a catalytic compound produced by cells which helps to promote some specific metabolic activity such as oxidation. As PROTEINS, most enzymes act only in dilute solution and within a narrow range of pH and temperature.

Eocene epoch a geological epoch within the TERTIARY PERIOD (about 55-38 million years ago).

Era a major geological time interval. From earliest to most recent, the sequence of eras is PRECAMBRIAN, PALEOZOIC, MESOZOIC and CENOZOIC.

Eukaryote an organism with cells with a discrete nucleus separated from the CYTOPLASM by a membrane and with distinct cytoplasmic organelles. Eukaryotes evolved from nucleus-lacking PROKARYOTES during the PRECAMBRIAN.

Facies general appearance of sedimentary rocks or fossils.

Family a rank in the taxonomic hierarchy coming between ORDER and tribe or GENUS.

Fauna the animals *in toto* of a given area.

Fertilization the union of male and female GAMETES in sexual reproduction.

Fission cell division splitting into two or more parts; this is a typical form of ASEXUAL REPRODUCTION.

Fitness a measure of the degree to which a given genetic type succeeds in reproducing itself. An organism's fitness is related to its ecological adaptedness and sexual proficiency.

Flora the plants *in toto* of a given area.

Foraminifera protozoa, usually marine, with perforated shells of chitinous, calcareous, or siliceous material. The shells of millions of foraminifera form the main component of chalk.

Gamete a male or female reproductive cell, usually with half the normal number of CHROMOSOMES so that at their union (FERTILIZATION) the resulting zygote has a full complement of chromosomes.

Gene flow interchange of genetic factors between and within populations as a result of emigration and immigration of individuals.

Genes the units of inheritance which are transmitted from generation to generation and control the development of an individual.

Genetic drift the occurrence of random changes, irrespective of selection and MUTATION, in the genetic make-up of small isolated populations.

Geneticist the biologist studying the inheritance of characters between generations.

Genetic variation the differences in genetic make-up between individuals of the same species (eg blue and brown eyes). It is the differential success of these variants that results in the process of NATURAL SELECTION.

Genotype the genetic makeup of an individual at a particular gene locus.

Genus a low taxonomic rank lying between FAMILY and SPECIES which consists of one or several closely related and morphologically similar species.

Golgi body cytoplasmic cell structure involved in the production of PROTEINS.

Gondwanaland the large Paleozoic continent consisting of South America, Africa, India, Australia and Antarctica.

Gradient a regular increase or decrease in any factor (eg humidity) in time or space.

Graptolites extinct marine organisms classified as Protochordates. Their horny exoskeletons are among the most common of Paleozoic fossils.

Gymnosperms plants, such as conifers, with seeds not enclosed in a true ovary. These evolved earlier than the flowering plants (ANGIOSPERMS).

Gypsum a mineral consisting of calcium sulfate.

Habitat the specific place and type of local environment that an organism lives in.

Half life a measure of the rate of radioactive decay: the time taken for the level of radioactivity to decrease by a half.

Haploid having only a single set of CHROMOSOMES: most GAMETES are haploid; cf DIPLOID.

Herbivore a plant-eating animal.

Heritability loosely, the capacity to be inherited; or more precisely, the component of an organism's total variability which is inherited in an additive fashion (ie in such a way that the effects of individual GENES add together).

Heterozygous having different genetic factors (ALLELES) at the corresponding locus of a CHROMOSOME pair (as opposed to HOMOZYGOUS).

Holism the belief that any complete system is more than the sum total of its component parts (as opposed to reductionism).

Hominid a man-like animal belonging to the evolutionary line leading to man.

Homology (homolog) a character in two or more TAXA that can be traced back to a single character in the common ancestor of those taxa. The character need not necessarily perform the same function in each case (eg the pentadactyl limb which serves as limb, wing or flipper).

Homozygous having the same genetic factors (ALLELES) at the corresponding locus of a CHROMOSOME pair (as opposed to HETEROZYGOUS).

Ichnology the study of TRACE FOSSILS and of recent animal traces.

Ichnospectrum the whole range of TRACE FOSSILS through geological time.

Invertebrate an animal without a bony or cartilaginous backbone.

Isotope any of two or more forms of a chemical element with differing numbers of neutrons in the atom. Many isotopes are radioactive (ie unstable).

Jurassic period a geological period of the MESOZOIC ERA (about 213-144 million years ago). This period was marked by a warm and stable climate and by the first appearance of birds.

Kin selection selection acting on an individual in favor of the survival not necessarily of that individual but of its relatives (which carry the same GENES). An

example is the ALTRUISM between genetically related members of a social insect colony.

Laurasia the northern land mass in the MESOZOIC consisting of North America, Greenland, Europe and Asia (except India).

Life form the product of all interacting life processes, both genetic and environmental.

Limestone a rock composed of calcium carbonate, usually derived from the skeletons of marine organisms.

Lineage the organisms in a single line of descent.

Lipids a group of fatty organic compounds, which, along with PROTEINS and carbohydrates, is one of the main organic components of living cells.

Lycopod tree-like club mosses.

Macroevolution evolutionary processes and phenomena above the species level, such as the origin of new higher TAXA, and evolutionary trends.

Mammal a member of the vertebrate class Mammalia, which consists of animals which produce milk with which they nurse their young.

Meiosis the process of cell division by which the number of CHROMOSOMES is halved; this consists of one chromosome duplication event and two successive cell divisions so that one diploid cell produces four HAPLOID daughters; cf MITOSIS.

Mesozoic era the geological era (about 248-65 million years ago), consisting of the CRETACEOUS, JURASSIC and TRIASSIC PERIODS.

Microbiota microscopic soil organisms not visible with the aid of a hand lens.

Microfossil a microscopic fossil.

Microhabitat a specialized, small and very local habitat/environment that can be exploited by small organisms.

Micropaleontology the study of microscopic fossils.

Migration the movement of organisms from one location to another. This may be periodic or seasonal (eg bird migration) or merely in response to local hardship.

Miocene epoch a geological epoch (about 25-5 million years ago) within the TERTIARY PERIOD.

Mitosis the process by which a single cell produces two daughter cells identical to itself by duplication and splitting of its CHROMOSOME complement and by fission; cf MEIOSIS.

Monotreme a mammal of the primitive order Monotremata, which comprises the platypus and echidnas. They are the only egg-laying mammals.

Morphology the structure and form, especially external, of an organism.

mRNA messenger ribonucleic acid. The genetic material which acts as an intermediate between the DNA in the nucleus and the sites of protein production in the cytoplasm. See RNA.

Mutation a sudden change in the genetic material. This most often involves the alteration of a single GENE but may also entail large scale changes such as alterations in chromosome number.

Nanoplankton minute plankton, including algae, BACTERIA and protozoa.

Natural selection the process whereby individuals with the most appropriate

ADAPTATIONS are more successful than others, and hence survive to produce more offspring. To the extent that the successful traits are heritable (genetic) they will therefore spread in the population.

Natural theology the synthesis of science and theology prevalent from the time of the Renaissance until the late 18th century, such that there was no conflict between the two. Nature, it was thought, provided convincing proof of the existence of a Supreme Being.

Nature philosophy a Romantic late 18th and early 19th century movement (*Naturphilosophie*) in German biology; a search for the unification of all natural phenomena through various transcendental and developmental beliefs.

Neo-Darwinism the modern theory of evolution which combines both NATURAL SELECTION and genetics.

Neo-Lamarckism the theory that characters acquired by organisms in response to environmental factors are assimilated into their genetic make-up and thence passed on to offspring. No firm evidence for this has ever been advanced.

Niche the role of a species within the community, defined in terms of its lifestyle (eg food, predators, competitors, and other resource requirements).

Nodule a small lump or protuberance. This may apply to a small aggregation of cells or to a lump of rock.

Nomogenesis a term for a postulated intrinsic perfecting principle which acts as a force in evolution.

Nucleic acid a molecule consisting of long chains of sugars, phosphate groups, and purine and pyrimidine BASES. An example is DNA which forms the basic hereditary material of cells and whose molecules have the ability to replicate themselves at cell division.

Nucleotide the structural unit of a NUCLEIC ACID; an ester of a nucleoside (pentose sugar linked to a purine or pyrimidine base) and phosphoric acid.

Oligocene epoch a geological epoch within the TERTIARY PERIOD (about 38-25 million years ago).

Oligonucleotide a group of a small number of NUCLEOTIDES bound together.

Ontogeny the process of growth and development to maturity of an individual.

Order a rank in the taxonomic hierarchy coming between CLASS and FAMILY.

Ordovician period a geological period in the PALEOZOIC (about 505-438 million years ago). This was marked by considerable land submergence.

Organism an entity capable of carrying out all life functions.

Paleobiogeography the study of the geographical distribution of fossil floras and faunas.

Paleobotany the study of fossil plants.

Paleocene epoch a geological epoch within the TERTIARY PERIOD (about 65-55 million years ago).

Paleoclimate the climate during periods of the geological past.

Paleoecology the study of the ecology of fossil communities.

Paleontology the study of fossils.

Paleozoic era a geological era (about 590-248 million years ago) consisting of the CAMBRIAN, ORDOVICIAN, SILURIAN, DEVONIAN, CARBONIFEROUS and PERMIAN PERIODS.

Pangaea the postulated former supercontinent, formed about 240 million years ago, and supposedly composed of all the continental crust of the earth. It later fragmented into LAURASIA and GONDWANALAND.

Pangenesis a theory of heredity proposed by Darwin in which somatic cells of the body were supposed to give off particulate "gemmules" which aggregate in the germ cells. These gemmules, which can be influenced by the environment and by the activity of the organs originally containing them, are inherited. This is erroneous.

Parallel evolution the maintenance of constant differences in characters through evolution between two unrelated lines. Alternatively, the independent acquisition of similar derived character states, evolved by related TAXA from a common ancestral condition.

Paraphyletic group a taxonomic group of descendants of a single ancestral TAXON but not containing all the descendants.

Peat an accumulation of unconsolidated and partially decomposed plant matter; characteristic of many waterlogged habitats.

Pelagic zone the upper part of a body of water, as opposed to the bottom substrate.

Peptide a compound of two or more amino acids joined by peptide bonds.

Permian period a geological period marking the end of the PALEOZOIC ERA (about 286-248 million years ago). This was marked by considerable mountain formation, glaciation in the Southern Hemisphere and the radiations of the primitive reptiles and modern insects.

Permineralization the process of fossilization whereby dissolved minerals infiltrate into the pore spaces of bones, shells, and other skeletal parts, and subsequently precipitate out of solution.

pH a measure of acidity/alkalinity.

Phenetics a system of classification based on overall similarity between organisms; characteristics are selected without regard to evolutionary history.

Phenotype the product of the interaction between the GENOTYPE and the environment; the observable characters of an organism.

Phosphorylated having undergone esterification with phosphoric acid.

Photodissociation the breakdown of molecules under the influence of light.

Photosynthesis the synthesis of organic compounds, primarily sugars, from carbon dioxide and water using sunlight as the source of energy, and chlorophyll, or some other related pigment, for trapping the light energy.

Phylogeny the evolutionary history or ancestry of a group of organisms. A classification based on phylogeny is termed **holophyletic**.

Phylum a major taxonomic category coming below kingdom and above CLASS.

Physiology the study of the biological processes which occur within living organisms.

Placental mammals (Eutheria) mammals in which the fetus is connected to its mother's blood supply by a placenta. This allows the young to be relatively advanced at birth.

Pliocene epoch a geological epoch within the TERTIARY PERIOD (about 5-2 million years ago).

Polymerization the bonding together of two or more monomers to produce a polymer (substance containing repeated units).

Polymorphism the occurrence of more than one form of an individual within a single species; the coexistence of a number of distinct genetically determined forms in the same population, where the frequency of the rarest type is not maintained by mutation alone.

Polypeptide a chain of amino acids linked together by peptide bonds.

Polyphyletic group a taxonomic group whose members are derived from more than one ancestral stock.

Polyploidy the state of having more than two sets of homologous CHROMOSOMES.

Polysaccharide a carbohydrate composed of many monosaccharide units linked together.

Population a more or less separate (discrete) group of organisms of the same species within a given biotic community.

Prebiotic reaction pertaining to reactions occurring under the conditions which existed before the origin of life.

Precambrian era all geological time prior to the beginning of the PALEOZOIC ERA, ending about 590 million years ago.

Predator an organism that preys on other organisms.

Prokaryote an organism whose cells have genetic material in the form of simple filaments of DNA not separated from the CYTOPLASM by a membrane. BACTERIA and blue-green algae are prokaryotes; cf EUKARYOTE.

Protein a complex organic compound composed of numerous amino acids joined together by peptide linkages, forming one or more folded chains. The sequence of amino acids is peculiar to a particular protein.

Protein clock a means of measuring the time since species shared a common ancestor, based on the assumption that the substitution of an amino acid at any place in a PROTEIN chain is likely to take place at random.

Protoplasm the living material of a cell, consisting of a complex colloidal mixture of PROTEINS, LIPIDS, carbohydrates, water, etc.

Psilophyte a member of the plant phylum Psilophyta; pteridophyte plants of which only one order is living today. They are distinguished from other vascular plants by having a rootless sporophyte.

Pteridosperm a seed-fern.

Race a taxonomic division subordinate to subspecies which links populations with similar distinct characteristics.

Radical a stable group of atoms forming part of the molecule of a number of compounds, both organic and inorganic.

Radiometry quantitative chemical analysis based on the measurement of disintegration rates of radioactive compounds; used for calculating the ages of objects/material samples.

Recessiveness the condition wherein a particular trait is only expressed phenotypically when the GENES (ALLELES) determining it are in the HOMOZYGOUS state.

Regression statistically, the estimation of the relationship between one variable and one or more others. Otherwise, a reversal in the direction of evolution or a retreat of the sea from a land area.

Replication the formation of an identical copy or copies (eg DNA replication at cell division).

RNA ribonucleic acid. A long-chain, usually single-stranded, NUCLEIC ACID. During the growth and reproduction of cells DNA molecules act as templates for the building of RNA molecules known as messenger RNA (mRNA), which in turn act as templates on which PROTEIN molecules are built. Another form of RNA known as transfer RNA (tRNA) functions so as to gather specific amino acid molecules and transfer them to their appropriate places on the mRNA template, so building particular proteins.

Sandstone a sedimentary rock composed of fragments of broken rocks, with grain size ranging from 1/16mm to 2mm in diameter. Grains are bound together by a secondary cement or welded together by pressure.

Scavenger an organism that feeds on carrion, organic refuse and similar matter.

Sedimentology the study of sediments and sedimentary rocks, which are formed when particular material is deposited in a fluid medium.

Sexual selection an evolutionary mechanism whereby females select for mating only males with certain characteristics, or *vice versa*.

Shale a well-laminated sedimentary rock that splits easily along bedding planes.

Siliceous composed of, or containing, silicate (which is found in many skeletal remains).

Silurian period a geological period within the PALEOZOIC ERA (about 438-408 million years ago). This was marked by extensive shallow seas, the first terrestrial arthropods, the first jawed fishes and a major radiation of terrestrial plants.

Sociobiology the study of the biological basis of social behavior.

Speciation the process by which new species arise in evolution. Two major mechanisms have been proposed: **allopatric speciation**, in which reproductive isolation is attained by populations that are completely geographically separated; and **sympatric speciation**, in which the populations attaining reproductive isolation are not geographically separated but overlap in their distributions.

Species a taxonomic division coming between GENUS and SUBSPECIES. In general, a species is a group of organisms of similar structure which are able to interbreed and produce viable and fertile offspring.

Stasis in paleontology, this refers to the constancy of form shown by some fossils over long periods of time.

Stratigraphy the study of the composition, distribution, origin and succession of rock strata.

Stratum a layer of rock distinguishable from adjacent layers.

Stromatolite a laminated concentric structure formed of calcium carbonate and occurring in some PRECAMBRIAN sediments. These were probably formed by algae.

Subspecies a recognizable subpopulation of a single SPECIES, typically with a distinct geographical distribution.

Sympatric speciation see SPECIATION.

Taxon a taxonomic grouping of organisms, or the name applied to it.

Taxonomy the study of the classification of organisms. It is generally convenient to group together animals which share common features and which are also thought to have common descent. Each individual is thus a member of a hierarchical series of categories: individual − SPECIES − GENUS − FAMILY − ORDER − CLASS − PHYLUM − kingdom.

Teleost a fish belonging to the infraclass Teleostei, of the subclass Actinopterygii (the ray-finned fishes).

Tertiary period a geological period of the CENOZOIC ERA (about 65-2 million years ago), marked by the rise of modern mammals.

Tethys the sea which separated LAURASIA from GONDWANALAND following the break up of PANGAEA in the MESOZOIC (about 150 million years ago).

Therapsid a mammal-like reptile of the order Therapsida, subclass Synapsida, which appeared first in mid PERMIAN times and persisted until the end of the TRIASSIC.

Tommotian fauna the earliest skeletalized faunas in the lower CAMBRIAN. It includes a wide variety of small shelled fossils, many composed of calcium phosphate.

Trace fossil a sedimentary structure resulting from the activity of a living animal, eg, a trail, track or burrow. They are found in ancient sediments such as sandstone, shale or limestone.

Transcription the synthesis of mRNA from a DNA template. See RNA.

Translation the process in which the base-sequence of mRNA is sequentially read and translated into a particular amino acid chain during protein synthesis. See RNA.

Triassic period the first geological period of the MESOZOIC ERA (about 248-213 million years ago). It was marked by the rise of the reptiles and the first mammals.

tRNA transfer RNA. See RNA.

Type a specimen on which the formal definition of a SPECIES or SUBSPECIES is based. A "type species" is the species designated as the type of a genus or subgenus.

Ungulate a term referring to those mammals which have their feet modified as hooves. Most are large and totally herbivorous (eg deer, cattle, horses).

UV ultraviolet. This refers to short wavelength electromagnetic radiation (4-380nm).

Variation differences between individuals which may be caused either genetically or environmentally. The differential survival of genetic variants results in the process of NATURAL SELECTION.

Vascular plant a plant with an internal vascular system of conducting tubules.

Vertebrate an animal with a backbone, belonging to the Vertebrata, the major subphylum of the phylum Chordata.

Virus a member of the large group of infectious agents simply composed of a protein sheath and NUCLEIC ACID core.

INDEX

A **bold number** indicates a major section of the main text, following a heading. A single number in (parentheses) indicates that the animal name or subjects are to be found in a boxed feature and a double number in (parentheses) indicates that the animal name or subject are to be found in a spread special feature. *Italic* numbers refer to illustrations.